SUSTAINABLE WATER RESOURCE DEVELOPMENT USING COASTAL RESERVOIRS

SUSTAINABLE WATER RESOURCE DEVELOPMENT USING COASTAL RESERVOIRS

T.G. SITHARAM

SHU-QING YANG

ROGER FALCONER

MUTTUCUMARU SIVAKUMAR

BRIAN JONES

SREEVALSA KOLATHAYAR

LIM SINPOH

ELSEVIER

Butterworth-Heinemann
An imprint of Elsevier

Butterworth-Heinemann is an imprint of Elsevier
The Boulevard, Langford Lane, Kidlington, Oxford OX5 1GB, United Kingdom
50 Hampshire Street, 5th Floor, Cambridge, MA 02139, United States

Library of Congress Cataloging-in-Publication Data
A catalog record for this book is available from the Library of Congress

British Library Cataloguing-in-Publication Data
A catalogue record for this book is available from the British Library

ISBN: 978-0-12-818002-0

For information on all Butterworth-Heinemann publications visit our website at https://www.elsevier.com/books-and-journals

Publisher: Matthew Deans
Editorial Project Manager: Isabella C. Silva
Production Project Manager: Nirmala Arumugam
Cover Designer: Christian J. Bilbow

Typeset by TNQ Technologies

Contents

Contributors

Reza Ahmadian
Hydro-environmental Research Centre, Cardiff University, Cardiff, United Kingdom

Simon Beecham
Natural and Built Environments Research Centre, University of South Australia, Adelaide, SA, Australia

Roger A. Falconer
Hydro-environmental Research Centre, Cardiff University, Cardiff, United Kingdom

Yuliang Gu
National Engineer Research Center of Urban Water Resources (South), Shanghai, China; Shanghai National Engineer Research Center of Urban Water Resources Co., Ltd., Shanghai, China

Bin Guo
Hydro-environmental Research Centre, Cardiff University, Cardiff, United Kingdom

Brian G. Jones
School of Earth, Atmospheric and Life Sciences, University of Wollongong, Wollongong, NSW, Australia

Sreevalsa Kolathayar
Department of Civil Engineering, National Institute of Technology Karnataka, Surathkal, Karnataka, India

Prerna Kumari
Department of Applied Mechanics and Hydraulics, National Institute of Technology Karnataka, Surathkal, Mangaluru, Karnataka, India

Sin Poh Lim
Global Water Consultants Sdn Bhd, Kuala Lumpur, Malaysia

Jianli Liu
Centre for Coastal Reservoir Research, University of Wollongong, Wollongong, NSW, Australia

Md Nasir Mohd Noh
National Hydraulic Research Institute of Malaysia (NAHRIM), Malaysia

H. Ramesh
Department of Applied Mechanics and Hydraulics, National Institute of Technology Karnataka, Surathkal, Mangaluru, Karnataka, India

T.G. Sitharam
Department of Civil Engineering, Indian Institute of Science, Bangalore, Karnataka, India; IIT Guwahati, Guwahati, Assam, India

Muttucumaru Sivakumar
School of Civil, Mining and Environmental Engineering, University of Wollongong, Wollongong, NSW, Australia

Yew Chong Tan
Formerly from Ministry of Energy, Green Technology and Water, Putrajaya, Malaysia

Fang Yenn Teo
University of Nottingham Malaysia, Semenyih, Malaysia

Cai e Wu
Shanghai Investigation Design & Research Institute Co., Ltd (SIDRI), Shanghai, China

Shu-Qing Yang
School of Civil, Mining and Environmental Engineering, University of Wollongong, Wollongong, NSW, Australia

Jian Zhong Yuan
Shanghai Investigation Design & Research Institute Co., Ltd (SIDRI), Shanghai, China

CHAPTER 1

Introduction to IACRR and coastal reservoirs (CR)

Muttucumaru Sivakumar[1], Shu-Qing Yang[1], T.G. Sitharam[2,3]
[1]School of Civil, Mining and Environmental Engineering, University of Wollongong, Wollongong, NSW, Australia; [2]Department of Civil Engineering, Indian Institute of Science, Bangalore, Karnataka, India; [3]IIT Guwahati, Guwahati, Assam, India

This book brings together coastal reservoir (CR) experts' contributions on some of the key issues relating to CRs in the world. This is a milestone produced by the International Association for Coastal Reservoir Research (IACRR; www.iacrr.org), which was officially inaugurated in 2017 in Kuala Lumpur, Malaysia. The IACRR was jointly founded by its members from China, India, the United Kingdom, Malaysia, the Netherlands and Australia. Now its influence has been expended to other countries such as Indonesia, Sri Lanka, South Africa and the United States. CR is a technology that develops river water in the sea near a river mouth; it is a paradigm shift in water resources development from storing water in mountainous dams to storing freshwater by the coast, from upstream water management to downstream water management. This technology can convert floodwater into valuable water resources closer to the coastal population centres where more than 50% of global population live. In the world, the run-off lost to the sea is approximated as 45,500 km^3/year, whereas the 7 billion people's drinking water is only 380 km^3/year, or 0.8% of the run-off.

The IACRR is the platform for sharing of knowledge and experience to ensure successful CR implementation worldwide. The IACRR's vision is to supply sufficient, high-quality and affordable water to the world with minimum environmental/social impact. The IACRR, different from other groups, believes that the world is not running out of water, but water, especially floodwater, is running out of our river mouths. The IACRR treats freshwater lost to the sea as precious water resources, similar to the existing scientific definition of water resources: surface water and groundwater. The IACRR is dedicated to advancing all aspects of CR development and promoting the sustainable development and management of freshwater otherwise lost to the sea.

Sustainable Water Resource Development Using Coastal Reservoirs
ISBN 978-0-12-818002-0
https://doi.org/10.1016/B978-0-12-818002-0.00001-0

After its inauguration, the IACRR is expanding quickly. The first international workshop on coastal reservoirs was held at the University of Wollongong. Subsequently many activities have been organized such as the first international conference on coastal reservoirs to be held in Hohai University, China soon. A number of CR workshops have been conducted in Australia, China, India, Malaysia and Taiwan (China). Many CR special sessions have been carried on such as those in the 37th, 38th IAHR world congresses, the 7th IWA-ASPIRE and CoastLab 2020. Many editorial boards of international journals and water related associations such as IAHR and IWA have paid great attention to the new technology; special CR issues have been published in IAHR's magazines—*Hydrolink*, *Journal of Sustainable Urbanization, Planning and Progress*, *Journal of Coastal and Ocean Engineering*. IACRR's voice has reached high-level water decision-makers such as Australian government, Malaysian government and Indian government, even UN-Water. Many funding agencies start to support CR research, and our members have been financially supported by Natural Foundation of Science and Technology, China, and Scheme for Promotion of Academic and Research Collaboration (SPARC), India. It is certain that in the near future, more R&D activities will be supported by more funding agencies and water industry. We appeal that more governments pay attention to our members' voice. It is certain that CR technology will be helpful to achieve SDG6, the sustainable development goal for water supply and sanitation in UN 2030 agenda.

The IACRR's team members have gradually changed the world's water storage map. Our Shanghai team has successfully secured the world largest city's water supply using the second-generation coastal reservoirs in the Yangtze estuary, which has minimum environmental/social impacts, and only highest-quality water is stored. IACRR's UK team members involved in the R&D works for Cardiff Bay coastal reservoirs, which has regenerated the coastal city. Our Netherlands team members have conducted research works for many coastal reservoirs in the world. Our Malaysian and Indian teams are investigating the feasibility of CRs for their local applications with government's support and endorsement. It is timely needed to exchange/share CR design experience. It is important for IACRR to coordinate the CR applications worldwide, to feedback the public and government's demand or comments to its members; thus, better and better CR designs/constructions/management can be achieved globally. It is why this book should be published. Its readers could be high-level government officers, university researchers and students, water engineers and water lovers.

Five months after IACRR's inauguration, the first international CR workshop was held in the University of Wollongong, Australia, in January 2018. Its theme is 'Innovative solutions to UN's Sustainable Development Goal 6', the workshop attracted about 100 delegates from the world including government decision-makers, parliament members, consultant/construction engineers, university researchers and even royal family members. Its water leaders' forum provided a good opportunity for water leaders to share their vision on CR technology, and the leaders endorsed 'WOLLONGONG DECLARATION ON WATER'. It says that 'A coastal reservoir is a unique structure, constructed at an estuary, gulf, bay or in the sea (i.e. the point where a river meets the sea) to store a portion of the excess water during times of floods. There are already CRs in operation in various countries, such as: the Netherlands, Singapore, China (Shanghai), South Korea, Hong Kong, India, etc. and the practice has proven to be beneficial. A coastal reservoir has many advantages such as: minimal land acquisition problems, no land submergence and no habitat or forest submergence – as occurs for in inland dams. The inlet and outlets can be designed in such a way that only good quality water enters the reservoir, with very little stagnation of impounded water. One can adopt very innovative designs, such as smart sensor networks, to efficiently operate such coastal reservoirs.' 'We appeal to the relevant governments, funding agencies, and the public to strongly support and adopt this promising technology in order to achieve SDG6. We, the members of the Water Leaders Forum at the International Workshop on Coastal Reservoirs, would like to confirm our full support for the IACRR Declaration on Water. We strongly believe that the implementation of this IACRR Declaration will help to attain and deliver the Sustainable Development Goals of the United Nations in the area of Water, especially in developing countries and also those developed countries near to the coast'.

Among IACRR's activities, it is worthwhile to note International Workshop for Coastal Reservoirs in Earthquake Active Areas in National Cheng Kung University, Taiwan in May 2019. Its theme is 'From Water Shortage to Water Security through Storage in Downstream Coastal Reservoirs'. Again, this workshop attracted about 100 delegators from earthquake active countries, including Taiwan water minister. It triggers the high demand from these countries for coastal reservoir development, because SDG6 becomes more challenging for earthquake active areas such as Taiwan, which is blessed by plentiful rainfall, on average 2.5 m/year, tripled the world average. All delegators agree that among all natural disasters, earthquakes threaten the largest number of people. In 2015, the number of people living in seismic areas was up to 2.7 billion (e.g. Pacific Ring of Fire). The mountainous dams in upstream become 'sword of Damocles' to the downstream people, and most of them live in the coast. All delegators appeal governments in earthquake active areas, funding agencies and the public to pay attention to this promising technology to SDG6. They strongly believe that CR technology can act as a barrier to protect the coast against natural hazards such as the tsunami and even in a worse disastrous event of an earthquake. The CR technology has no threat to human habitat unlike dams. CRs have the potential to create a new freshwater ecosystem and can potentially be developed into largest freshwater aquaculture farm. The existing inland dams are swords hanging over the head, whereas CRs can provide ample water for drinking and even irrigation purposes with zero risk in earthquake events. It also provides opportunities for food—energy—water nexus, as well as water sports in the reservoir and tourism attractions, even new urban generation. The presence of coastal reservoir will pave the way for improvement in the quality of groundwater (saline water to freshwater) near the coast. In short, a coastal reservoir has the potential to transform neglected and polluted coastal area into a sustainable freshwater township.

It is obvious that all of these agreements achieved in these workshops should be sent to coastal communities for their own welfare. We should not hide the conflict between CR technology and the desalination technology. It is true that most governments in the world are considering the desalination technology as their future water solution. The CR technology provides one more option for coastal communities to choose. Therefore, the publication of this book would be very useful for the debate. We believe that the nature has built a giant desalination plant for the human, i.e. the solar energy is used to evaporate seawater, and then the vapour is condensed and precipitation occurs, run-off forms and life thrives in this planet. Any artificial desalination plant just likes a burning candle under sunshine, especially for those in run-off-rich regions.

Currently, the IACRR has several local chapters. Its Australian chapter is investigating CR's feasibility for the driest inhabited continent in the world. Australia only uses 20 km^3/year, or 4.5% of its annual average run-off. Malaysian chapter is investigating CR feasibility for its islands and capital cities. Malaysia only uses 2.7% of its available water resources. Similarly, Indian and China chapters are investigating CR technology to secure their 2.8 billion people's water demand.

This book has 14 chapters; most of these papers were present and discussed in Wollongong CR workshop. The authors have revised their original papers again and again after considering other experts' comments. The authors represent IACRR members from Europe, Asia and Australia continents. They are composed of CR planers, designers, researchers and decision-makers. This book covers the CR concepts, design principles, CR examples and future applications; in other words, it contains almost every aspects of CR technology, from past experience to future application, from water quality to water quantity, from water engineering to geotechnic engineering, from climate change to remote sensing monitoring and from environmental impacts to social impacts. We are confident that the publication of this book can clarify many queries in those who first knows the CR technology.

Chapter 2 discusses the most important issue of CR technology, i.e. water quality. Different from mountainous dams that collect rainwater from a virgin catchment, all CRs are located in the end of river, which has the potential to collect and store all containments from a basin, or CRs are likely to become a wastewater reservoir. Most of the existing CRs in the world have suffered this problem, and the Sihwa Lake in South Korea is one of the examples. The past experience of CRs is summarized, and new methods for high-quality water are developed. Some critical parameters for CRs are suggested including CR's location selection, size, shape and layout plan. It is recommended that CR's storage capacity needs to match its targeted water demand, and shallow water bodies should be excluded during site selection. Its inlet(s), outlet(s) and pumping stations' design should minimize stagnant water body. It is interesting to note that the chapter discusses how to improve the water quality in existing CRs in the world by using the suggested design methods.

Chapter 3 continues the discussion of water quality contaminants, which is the worst enemy of coastal reservoirs. The authors highlight the importance of catchment management estimation of conventional pollutant loads and awareness of emerging pollutants, and Australian experience is introduced.

Chapter 4 presents the geotechnical considerations in CRs' construction. Different from high dams whose design mainly depends on structural engineers' skills, CRs' barrier generally is less than 30 m high from the top to the bottom, and the water pressure is balanced by the seawater, leading smaller net water pressure force on both sides of seawall; thus, geotechnical engineers generally play a leading role in CRs' development. Different from earth dams, the CRs' geotechnical consideration includes seawater intrusion from the seabed and surfing from the top, the seawall could be constructed by giant geotubes and caissons. Therefore, many opportunities and challenges come from CRs' barrier for geotechnical experts to deal with.

Chapter 5 demonstrates the successful example of CRs' application for SDG6 in Shanghai, China. In the end of the 20th century, water experts from the United Nations and World Bank asserted that, in the 21st century, Shanghai would be one of six megacities in the world with severe water shortage induced by water pollution, and this conclusion was supported by Chinese water ministry. IACRR's Shanghai team made their attempts to apply the CR technology to develop freshwater in the Yangtze estuary. The Qingcaosha reservoir by far is the largest CR in the world for water supply purpose, and it benefits large population and represents the most cost-effective investment among other water solutions such as desalination plants, wastewater recycling plants and water diversion from remote areas. The authors of the chapter are the master planners, chief engineers and managers of Qingcaosha CR from its early stage research, measurement, construction and till today water quality management. From its concept to construction, a lot of scientific and technical difficulties have been overcome. This chapter elaborates the Shanghai story from a CR idea to strategic layout, from first-generation CR to the second-generation CR and its logic linkage with China's old Dujiangyan project. Shanghai team's achievement and experience are worthwhile to be shared with the international communities who are looking for water. Yangtze River is the third largest river in the world; its estuary is most difficult to construct CRs due to high speed currents, complex seawater intrusion, typhoons and soft soil seabed, river regime stabilization and water source protection. It is predictable that in the near future, no any CRs in the world are more difficult to develop compared with Qingcaosha CR.

Chapter 6 is also contributed by the IACRR's Shanghai team, the chief designers. The authors reviewed the water problems in Shanghai from 1900 to 2000 and concluded that only the CR solution is appropriate for Shanghai. The main function of the Shanghai coastal reservoirs is to store freshwater during low salinity and high water quality for use during the dry season. Yangtze River is difficult to apply CR technology because it suffers typhoons, severe saltwater intrusion caused by high tides, oil pollution accidents by busy navigation and very high agricultural/industrial activities in upstream where the total population is equivalent to all EU people.

The authors share their experience in designing dikes, hydraulic structures, intake and other water supply facilities to ensure that water quality meets the requirements of the design standards all the time. Other experience is also included like site selection, arrangement of sluice and pump stations, scheduling scheme, gap closure, water quality protection, reservoir eutrophication prevention and control technology, hydraulic fill dam in soft soil foundation, closure gap setting, protection and closure, permeable foundation pit maintenance and foundation treatment. The experience from the Shanghai coastal reservoirs can provide a technical reference for similar coastal reservoir in other parts of the world.

Chapter 7 demonstrates how CR technology can supply sufficient, high-quality and affordable water to Australia with minimum environmental/social impacts. The driest country is also a leader in urbanization, and now 86% of Australia's population lived in cities in 2018, about 85% of Australians living within 50 km of the sea. Obviously, once the coastal water demand is secured, the inland water stress can be also mitigated. The authors from the University of Wollongong analyzed the data in every capital city; they found that Adelaide is the water-richest city due to the MDRB, the largest river in Australia. The second water-richest city is Melbourne where Australia water tower—the Snowy Mountain feeds the rivers and lakes in the east side of the city. The Greater Sydney is blessed by two large rivers in its north (Hunter River) and south (Shoalhaven River); once the run-off to the sea is developed, it is enough to feed the other three to four Sydney cities. The Brisbane has the highest rainfall; once a CR at Brisbane River mouth is constructed, the city's water demand is fully met and its flood disaster like 2011 big flood could be greatly mitigated. Their research outcome interestingly shows that Perth's water shortage comes from its storage shortage; its largest dam can only collect water from 0.8% of its Swan river basin. The researchers also propose water diversion from its existing dams to the inland regions once the coastal water demand is secured; thus, inland regions' water supply can be increased.

Chapter 8 extends CR's benefits to urban regeneration and renewable energy generation, and a vivid example is used to demonstrate the feasibility. The successful CR example for such an application is first presented, followed by numerical model's results that support positively this possibility. The authors considered two examples, including the Cardiff Bay Barrage project and the West Somerset Lagoon proposal. For Cardiff Bay, a barrage was constructed across the mouth of the bay, creating a coastal reservoir and reconnecting the estuarine waterfront with Wales' capital city, namely

Cardiff. The project was completed in 2000, and the reservoir has been highly successful in regenerating the region. For the other project, the authors suggested to build a large lagoon (or coastally attached reservoir) in the Bristol Channel to create tidal energy.

Chapter 9 discusses the climate change and coastal infrastructures. The CRs' life span is expected to be around 100 years; thus, the long-term climate change is important for all CR designers to consider. Especially, current climate models show that, in future, water availability is becoming less predictable in many places. Floods and droughts are more frequent and more severe, which makes water resources development more difficult; subsequently, water scarcity may be more common. The author's analysis shows that, since 1950, most of eastern and southwestern Australia has experienced substantial rainfall declines. From 1950 to 2005, extreme daily rainfall intensity and frequency has increased in northwestern and central Australia but decreased in the southeast and southwest and along the central east coast. His model outcome indicated changes in design rainfall values under nonstationary conditions. Mean, standard deviation and skewness coefficient are key parameters with which future uncertainties are associated, and the uncertainties increase as the project design life increases.

Chapter 10 focuses on CRs' application in India. The authors' analysis shows that India receives an average annual precipitation of 4000 km^3/year, and most of it from a 4-month monsoon period. The current water storage capacity of India is 350 km^3, less than 10% of the annual precipitation. Indian population over the decades has increased so much; the density of population and dwindling forest coverage are not allowing creation of any new dams. The original Indian proposal of interlinking of rivers to overcome the ill effects of variable rainfall is practically dead and impossible to implement under the changed circumstances and huge cost of the project. The CR solution fits India best as it develops floodwater without submerging lands. A concept of linked CRs along Indian coast is proposed, called as Sarovar mala around the Indian peninsula, which will be a sustainable solution for Indian water resources development specifically in the coastal areas.

Chapter 11 takes the Murray–Darling River basin (MDRB) as an example to apply the strategy of downstream water management in arid or semiarid regions. MDRB is the largest river basin in Australia, the driest country. Before European settlement, its lower lakes at the river mouth are blackish water bodies. In dry seasons, seawater could intrude upstream up to 250 km, which badly impacted agricultural irrigation. South Australia government decided to construct barriers to seal the lakes' outlets, and freshwater water coastal reservoir was formed in the 1930s. This is one of the earliest CRs for the purpose of agricultural development. However, the first-generation CR's design does not meet the criteria mentioned in Chapter 2 its salinity in the Millennium Droughts (2000—10) was too high to use, and the original design purpose was not achieved. The chapter suggests the second-generation CR design, and its feasibility was investigated using the MIKE-21 software. The outcome is surprising that the quality of water supply is improved and secured for the surrounding cities, even the ecosystem is improved once the suggested design is applied.

Chapter 12 introduces few new water storage projects in Malaysia. In its 2018 budget, the Malaysian government listed several water resource development projects emphasizing off-river storage (ORS), a downstream reservoir concept. This concept is gaining popularity after several successful ORS implemented projects, solving both quantity and quality problems that persist in the traditional approaches of Run-Off River Schemes or Regulated Dam systems. It has also been proven to be more economically, socially and environmentally friendly. Water quality is critical for downstream coastal reservoirs. Malaysia will install a series of real-time water quality monitoring stations upstream of the proposed CRs; thus, only selective good-quality water enters the reservoir. They also intended to construct a wetland at the upper part as a natural filtration system to improve the water quality. The authors recommended that the State Government of Malaysia consider the construction of CRs for major coastal cities in the country such as Johor Bahru, Melaka, Penang, Kuantan and Bintulu. Setting CRs as one of the requirements for new onshore, nearshore or reclamation land development at these areas can be one way to incentivize the development of CRs.

Chapter 13 discusses how to apply high technology for CR's design and how to get reliable nearshore bathymetry using remote sensing image; examples from Mangaluru coast, India, are provided to prove its accuracy. This technology has been used in Netravati CR design, and Indian government has endorsed the feasibility study report.

CHAPTER 2

Water storage and design considerations of coastal reservoirs

Shu-Qing Yang, Muttucumaru Sivakumar
School of Civil, Mining and Environmental Engineering, University of Wollongong, Wollongong, NSW, Australia

2.1 Introduction

The groundwater plays an important role for human survival in history. Hand-dug wells are present in every corner in the world. But this approach, while serves a small population, gradually became ineffective after urbanization which heavily depends on water supply from large-scale reservoirs. In the 1930s, the Hoover Dam across the Colorado River was the first high dam in the world which even conquered the US President Roosevelt as he confessed in his inauguration speech. Quickly, every country starts to construct the high dams in mountainous regions. Two notable associations are directly originated from the Hoover Dam: the World Commission on Dams and International Commission on Large Dams.

Fig. 2.1 shows the distribution of large inland dams constructed in the world, and Fig. 2.2 shows the dam construction in the 20th century. After 1980, the number of dams constructed was reduced year by year, because many negative impacts of dams on the environment have been discovered. For example, in 1976, the Sydney Catchment Authority constructed its last dam, the Tallowa Dam on the Shoalhaven River, New South Wales. With population growth, the government has proposed to construct more dams, but none of the government's proposals has been implemented due to public's opposition, like the Tillegra Dam proposal that has been cancelled three times since the 1960s. Other water supply solutions such as seawater desalination and wastewater recycling while still used in practice have large environmental footprints.

In the 1930s, the Netherlands and Australia constructed another type of water storages in seawater environment to contain freshwater, we call it as coastal reservoirs (CRs). The 32-km-long IJsselmeer Dam was constructed in 1932 in the Netherlands as shown in Fig. 2.3A, and the 1100 km^2 of

Sustainable Water Resource Development Using Coastal Reservoirs
ISBN 978-0-12-818002-0
https://doi.org/10.1016/B978-0-12-818002-0.00002-2

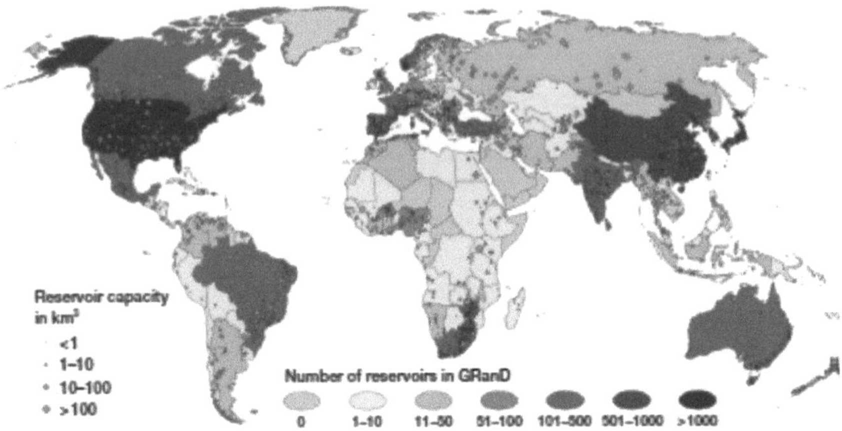

Figure 2.1 World large reservoirs' distribution by countries. *(Source: Global reservoir and dam (GRanD) Database. After Lehner, B., Reidy Liermann, C., Carmen, R., Vörösmarty, C., Fekete, B., Crouzet, P., Petra Döll Endejan, M., Frenken, K., Magome J., Nilsson, C., Robertson, J. C., Rödel, R., Sindorf, N., Wisser, D., 2011. High-resolution mapping of the world's reservoirs and dams for sustainable river-flow management. Frontiers in Ecology and the Environment 9 (9), 494–502. http://ds.doi.org/10.1890/100125.)*

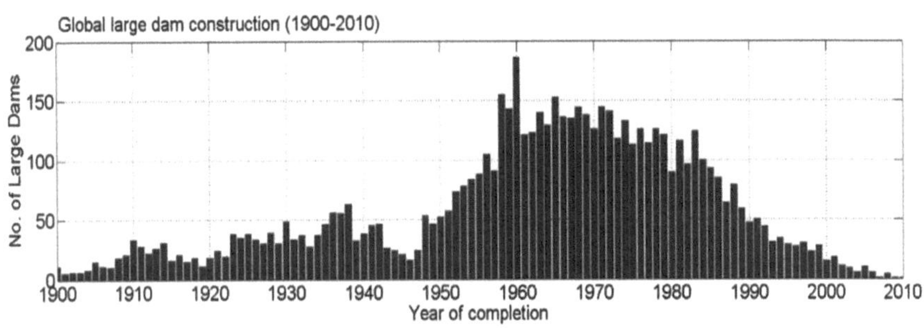

Figure 2.2 Global dam construction over the past 100 years. *(Source: Global reservoir and dam (GRanD) Database. After Gibbes B., Quiggin J., October 30, 2014. Dam hard: water storage is a historic headache for Australia. The Conversation. https://theconversation.com/dam-hard-water-storage-is-a-historic-headache-for-australia-33397.)*

reservoir became freshwater 5 years later after its closure. Almost at the same time, Australia, the driest inhabitated continent, made its first attempt to change Lake Alexandrina at the Murray–Darling River into a freshwater lake in the 1930s, as shown in Fig. 2.3B. A total of five barrages were constructed at the lake's outlets. Before the barrages were built, seawater intrusion during low flow could move 250-km upstream from its mouth

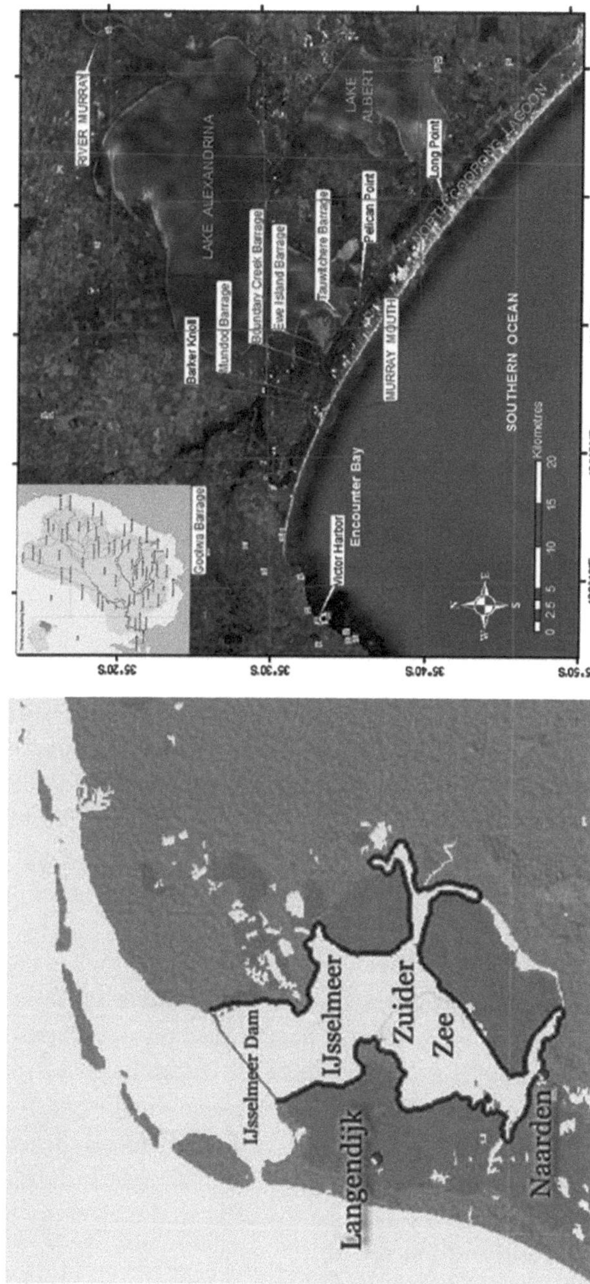

Figure 2.3 Coastal reservoirs in the Netherlands (A) and Australia (B).

(Yang, 2017), which had big impacts on irrigation as the farmers used the river water for agricultural activities. During the Australia Millennium Drought in 2000–09, the measured salinity in the lake was found being higher than seawater, i.e. 6000 EC in 2009 far above 800 EC, the acceptable drinking water guideline (Liu et al., 2018). This failed CR implies that its original design and size have not worked. More innovative design is needed.

In 2015, the United Nations' (UN) 193 members developed and endorsed 17 Sustainable Development Goals. Among them, Goal 6, to provide 'safe and affordable drinking water for all by 2030', is one of the most difficult or almost impossible to achieve, because currently about 40% of the global population are being affected by water scarcity, and this percentage is going to rise. A new and novel water solution is urgently needed for every government. This chapter mainly deals with why CRs may dominate future water supply and could be an effective solution for UN's SDG6. The quantity of available water, water quality and key design parameters are analyzed, and the possible environmental/social impacts are compared.

2.2 Coastal reservoir's definition and classification

A simple definition of CR is a freshwater reservoir inside seawater to capture freshwater run-off before being lost to the sea and mixes with salt water (Yang, 2004, 2016). CR's embankment will be situated in seawater, and preventing salt water mixing is a main consideration for its design. More strict definition could be in the following way: a reservoir inside a large water body stores fluid that is different from outside water in physical, chemical and biological properties intended to be used for a specific purpose. For example, the freshwater storage in seawater implies the salinity difference of inside reservoir water and outside seawater.

CRs can be classified into different forms based on their locations, dam types and design purpose as shown in Table 2.1. It can be seen that the first type of CRs is constructed in seawater, but the second type of CRs inside a lake, but the purpose is the same, i.e. water resources development, treatment of heavily polluted water and flood disaster mitigation. The details of the second type of CRs can be found from the literature by Wang et al. (2009) and Yang and Liu (2010).

The dam/dike could be constructed by different materials such as concrete, earth or soft dam and even logs. The trend of CR dam's development shows that more and more soft materials are used today, like

Table 2.1 Preliminary classification of coastal reservoirs.

	Description	Fluid inside reservoir	Large water body	Property difference
By location in the sea (first type)	Inside a river mouth (first generation)	Freshwater	Estuary	Salinity, TN, TP etc.
	Beside a river mouth in the sea (second generation)	Freshwater/ fluid for intended use	Seawater	Salinity, TN, TP etc.
	Outside a river mouth (offshore/ second generation)	Freshwater/ fluid for intended use	Seawater	Salinity, TN, TP etc.
In a lake in middle-lower river plain (second type).	For flood disaster mitigation and water resources development, wanted water flows to the reservoir, unwanted water bypasses the reservoir	Floodwater/ high quality	Fresh lake water	Turbidity, TN, TP, etc.
By dam types	Concrete dam	Wanted fluid	Lake/ estuary/ sea	Salinity, turbidity, etc.
	Earth dam	Wanted fluid	Lake/ estuary/ sea	Salinity, turbidity, etc.
	Soft dam	Wanted fluid	Lake/ estuary/ sea	Salinity, turbidity, etc.
Intended purpose	High-quality water supply	Freshwater	Lake/ estuary/ sea	Salinity, TN, TP
	Wastewater treatment for environmental protection	Highly polluted water	Lake/ estuary/ sea	Salinity, TN, TP
	Electricity/energy storage	Wanted water	Sea	Water level
	Recycling ships' ballast water in harbours	Ballast water	Sea	Biomass

rubber dams in Singapore's marina barrage, geotubes filled with soil/sand in Shanghai's Qingcaosha CR (Yang, 2015a,b).

The purpose of CR construction could be very broad, and Table 2.1 only lists some of them, namely water-supply CR, power-supply CR, CR for wastewater treatment and CR for recycling ballast water from ships in a harbour. Ballast water discharge from ships contains nonnative, nuisance, exotic species such as plants, animals, viruses and bacteria, which gives rise to serious ecological problems to local marine life. A ballast water CR can completely eliminate the problem caused by spread of invasive species. Of course, CR can also be used for flood control effectively.

Currently, many CRs exist around the world as listed in Table 2.2 (Liu et al., 2013). Except for the reservoirs on the Yangze estuary, all others are the first-generation CRs, i.e. a straight and short barrage creates the reservoir by enclosing a river mouth or a bay to store the incoming freshwater. As these CRs also collect all contaminants, generally these reservoirs may become wastewater reservoirs, such as the Sihwa Lake in South Korea (Bae et al., 2010). The worst enemy for any CR is the water quality of the source water, which requires innovative solutions.

Other shortcomings of first-generation CRs also include the following:

- They cut off connections between river and sea like sediment, ships' navigation and fish migration; thus, they have significant environmental impacts.
- The storage capacity of a CR is not able to match the designed water demands; it may be too big or too small. If it is too big, the evaporation loss will be high particularly if it is a shallow storage, and the stagnant water body may have adverse effect on its quality.
- The accumulative contamination (like salt, sediments and nutrients) in the reservoir may deteriorate its water quality.

In 1968, Plover Cove in Hong Kong became the first CR in the world for drinking water purposes. Most reservoirs in Singapore are CRs, such as Pandan, Kranji and Marina Bay. In China, Shanghai constructed the Baogang CR in 1985 and Chen Hang CR in 1992. Currently, its water supply mainly comes from the 550 GL Qingcaosha CR, which is by far the largest CR in the world for drinking water purposes. Its construction cost in 2011 was US$2.8 billion to supply 2600 GL/a for 12 million people.

The second-generation CRs should avoid the mentioned disadvantages with the following features (Fig. 2.4):

(1) Off-stream freshwater storage is located in the sea or estuarine space, rather than across the stream storage.

Table 2.2 World's existing coastal reservoirs.

Name	Catchment (km^2)	Dam length (km)	Capacity (GL)	Year completed	Construction cost (US$ B)	Country – River
Zuider Zee	170,000	33	5,600	1932	n.a.	Netherlands
Lakes Alexandrina and Albert	1.1×10^6	7.6	1,600	1940	n.a.	Australia – Murray
Plover Cove	46	2	230	1968	0.07	Hong Kong
Baogang/Shanghai	1.8×10^6	3.7	12	1985	n.a.	China – Yangtze
West Sea Barrage	20,000	7.8	2,700	1986	1.8	North Korea – Taedong
Cheng Hang/Shanghai	1.8×10^6	4.7	8.0	1992	n.a.	China – Yangtze
Sihwa	476	12.4	323	1994	0.3	South Korea
Yu Huan	170	1	64	1998	0.1	China – Zhejiang
Marina Barrage	110	0.35		2008	0.2	Singapore
Qingcaosha/Shanghai	1.8×10^6	43	550	2011	2.8	China – Yangtze
Saemanguem	330	33	530	2011	2.1	South Korea

n.a. = not available.

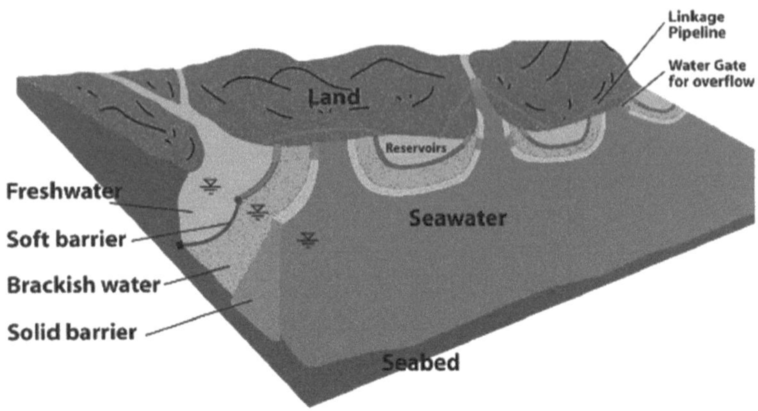

Figure 2.4 Scenarios for second-generation coastal reservoirs (Yang et al., 2001).

(2) The dam/dike is no longer short and straight, but long and curved to form a freshwater body.

(3) River water, sediment and ship can enter the sea without interruption except for a short period during discharging high-quality water to the CRs; thus the CR's adverse impacts on ecosystems are lowered to levels acceptable to the public and to governments.

(4) The CR can select high-quality water for storage and bypass polluted water into the sea. This is one of the most important features of the second-generation CR, i.e. supply of high-quality water.

(5) More flexibility in design, construction and operation of barrages for existing conditions as well as future changes in demand, quality standards, environmental expectations, etc.

Table 2.3 summarizes the differences between CRs and traditional inland reservoirs.

Fig. 2.4 shows the various scenarios of second-generation CRs in which the intake system is shown as an orange square, which can divert good-quality river water to the reservoir for storage. If water quality is not a major issue, both the estuary and sea space can be used for water storage (on the left). The interbasin water diversion can be achieved by linking CRs.

2.3 Design considerations of the second-generation coastal reservoirs

Selection criteria for CR site and its dike length. For an inland reservoir, like the Three Gorges Dam in China, the most important considerations are the dam site and its dam height, as both determine the

Table 2.3 Difference between inland reservoirs and coastal reservoirs.

Aspect	Inland reservoirs	First-generation coastal reservoirs	Second-generation coastal reservoirs
Water quality	Good (from protected catchments)	Poor (collect and store all contaminants)	Good (only collect clean water, bypass polluted water)
Water level	Variable water level, above sea level	Variable water level near sea level	Almost constant water level near sea level
Dam alignment	90 degrees to flow direction	90 degrees to flow direction	Small angle to flow direction
Dam site	Limited (narrow dam sites with)	Limited (only inside a river mouth)	Unlimited (inside or outside river mouth)
Dam design	Concrete, earth/rock to withstand high water pressures	Concrete, earth/rock to withstand low water pressure but with wave/tidal surges	Concrete, earth/rock to withstand low water pressure but with wave/tidal surge, with/without soft dam
Dam length	Short	Short	Long
Environmental impacts	High	Medium (obstruction to floodwater, fish, navigation)	Low
Seepage	By pressure difference	By density difference	By density difference
Pollutant	Land based	Land-based + seawater	Land-based + seawater
Emigrant cost	High	None	None
Water extraction	By gravity	By pump	By pump
Water catchment	10%–50%	100%	100%

storage capacity, total cost, etc.; others are less important, such as spillway, ship locks and hydropower stations. For a CR design, the most important considerations include the following:

(1) The intake of CRs should be located at an ideal place enabling to divert stable and high-quality water with minimum salinity at designed dry years

(2) The CR site's selection should consider its total construction cost, and the volume of enclosed water body should meet:

$$V_0/A = \text{maximum; and the total cost} = \text{minimum} \tag{2.1}$$

where V_0 = water volume of the reservoir at the normal water level and A = reservoir's surface area. For an offshore CR over a flat seabed, a circle embankment requires the minimum dike length. For a CR, its dike length is determined by the minimum cost by satisfying the area A in Eq. (2.1).

Seawater prevention and salinity estimation. For a CR as shown in Fig. 2.5, the salt is a major threat; its possible sources of water discharge (Q) and salinity (C) are (1) river flow (Q_1) whose salinity (C_1) is small but not zero, dependent on its catchment's characteristics; (2) seawater seepage (Q_2, C_2) from the reservoir bed and the embankment like ship locks; (3) overtopping (Q_3, C_3) from the dike driven by waves in extreme conditions; (4) the precipitation (Q_4, $C_4 = 0$) and $Q_4 = IA$ where I = annual precipitation rate.

The outgoing discharge and salinity are flow from the reservoir to the sea (Q_5, C) and water withdrawal (Q_6, C), and evaporation loss (Q_7, $C_7 = 0$) and $Q_7 = EA$ where E = annual evaporation rate.

The water volume V of the reservoir must meet the water balance equation:

$$\frac{dV}{dt} = Q_1 + Q_2 + Q_3 + Q_4 - Q_5 - Q_6 - Q_7 \tag{2.2}$$

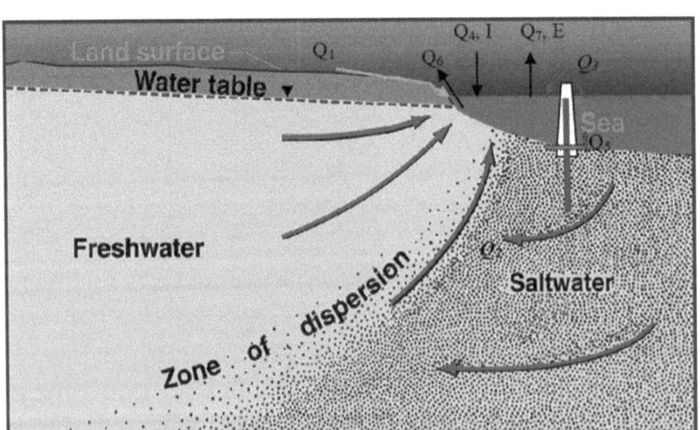

Figure 2.5 Interaction of seawater and freshwater in a coastal reservoir where the *red thick line (dark grey in print version)* is the impermeable wall and the *green dotted line (light grey in print version)* with an *arrow* in the dike is the drainage system.

The salt in the reservoir must also meet its mass conservation equation:

$$\frac{dVC}{dt} = Q_1C_1 + Q_2C_2 + Q_3C_3 - (Q_5 + Q_6)C \tag{2.3}$$

If we assume that the long-term average water volume is constant and Q_2 and Q_3 are negligible relative to Q_1, Eq. (2.2) gives:

$$Q_6 = Q_1 + Q_4 - Q_5 - Q_7 \tag{2.4}$$

Eq. (2.4) indicates that the evaporation loss may be an important factor in dry years ($Q_5 \approx 0$; $Q_4 \approx 0$). To reduce or neglect Q_7, one must satisfy Eq. (2.1), which suggests that a CR should be constructed in a deeper water body, and the shallow water region should be excluded for freshwater storage.

Eq. (2.3) can be simplified as follows:

$$V\frac{dC}{dt} = Q_1C_1 - (Q_5 + Q_6)C \tag{2.5}$$

By integrating Eq. (2.5) with respect to time t, the salinity of reservoir water can be obtained with the following form:

$$C = (C_0 - C_\infty)\exp\left(-\frac{Q_5 + Q_6}{V}t\right) + C_\infty \tag{2.6}$$

where

$$C_\infty = \frac{Q_1}{Q_5 + Q_6}C_1 \tag{2.7}$$

where C_0 = initial salinity in the reservoir, or the seawater's salinity can be assumed and C_∞ = the equilibrium salinity in the reservoir. From Eq. (2.6), one can estimate the required desalinization time for a CR.

In extreme dry year with $Q_4 = 0$, substituting Eq. (2.4) into Eq. (2.7), one has

$$C_\infty = \frac{Q_1}{Q_1 - EA}C_1 \tag{2.8}$$

Eq. (2.8) clearly indicates that if Q_1 = EA, the CR's salinity could be very high, and this was observed in 2009 in Alexandrina Lake, Australia, as mentioned previously. This is why Eq. (2.1) must be satisfied and the surface area A must be minimized to reduce the evaporation loss.

Fig. 2.5 shows the seawater intrusion Q_2 and Q_3 by seepage and overtopping flow. In dam engineering, the traditional methods are available and effective to eliminate the seepage Q_2, e.g. an impermeable layer inside the dike and seabed. The seawater has higher density than freshwater, or freshwater will appear always above the seawater; thus this chapter suggests that hydraulic gates of a CR should be located at lowest trench to discharge the saline water at the bottom. The near-bed drainage system should be installed along the dike; during low tide, when the reservoir water level is higher than the outside sea surface, these hydraulic gates and drainage system can flush the saline water (Q_5) out of the reservoir as shown in Fig. 2.5.

The aforementioned discussion outlines the main measures to prevent the seawater intrusion, which can be well controlled as long as the design is appropriate; only few CRs are failed because of high salinity in Australia and the Netherlands. CR's worst enemy is the water pollution in the river caused by both point and nonpoint sources; almost all existing CRs in Table 2.2 have the water quality/algal blooms problem. It is important to discuss how to design a CR for its high-quality water supply.

Land-based pollution prevention and wetland pretreatment. A good CR design should be able to select high-quality water for storage and bypass the unwanted/polluted water into the sea. This is achievable as shown in Fig. 2.4 using an intake in upstream. The intake is open to fill in the CR when high-quality water appears, but it is closed when the quality of river water does not satisfy the guideline values. The water quality is monitored along the river course by measuring sensors; experienced CR managers can easily judge when the intakes should be opened or closed.

It should be highlighted that the water quality in a river is a random parameter; its probability follows some distribution, like the log-normal distribution. In other words, the best quality water and the worst quality water have a small fraction (say 5%), and most river water's quality is between these two extremes (or 90%). The best quality water can be diverted to one of the CRs in Fig. 2.4, but the worst quality water can be also stored in another reservoir for treatment before discharging, so that the marine environment is protected. If needed, the CRs shown in Fig. 2.4 can be used for flood disaster mitigation also, i.e. before floods appear, the reservoirs need to be emptied to low tidal level, and the active capacity can be used for the peak flood.

In very large rivers such as Yangtze or Grande River, the best quality river water may be still not good enough; in such case, wetland pretreatment may be useful. Generally, the wetland can remove 50%–90% of nutrients in the river water (Yang, 2015b; Yang and Lin, 2011), an eco-agricultural wetland is specially recommended as the farmers can harvest the vegetation regularly, and the large arable land is still productive.

It is very important to have a bypass channel in all reservoirs to discharge unwanted/polluted water. If so, the failed CRs such as South Korean Sihwa Lake or others can improve its water quality significantly. This is the most important measure for CRs' water quality. For high-quality water supply, it is suggested that two main principles to follow for CR's design:

(1) The dike to enclose deeper water body as a reservoir must exclude the shallow water body.

(2) A bypass channel must be included in a reservoir.

Layout of hydraulic structures. The hydraulic structures for any inland dam include spillway, power station, ship locks, etc.; similarly for a CR, such structures include its inlet(s), outlet(s), pumping station(s), ship locks and so on. Its optimal layout must meet the following criteria:

(1) The minimum detention time or mathematically $V_0/Q_1 \rightarrow$ minimum

(2) The minimum stagnant water body

(3) The storage capacity matches the water demand, not too big or too small

In principle, the shape of CR, inlet/outlet locations and pumping stations have a joint effect on the hydrodynamics of the system; a good design can reduce the occurrences of algal blooms in the reservoir; partition walls can be used inside the reservoir to reduce its stagnant water body.

2.4 Learning from the past for the future

By following the design guidance mentioned earlier, the past CRs can be improved, and the preliminary proposal is shown in the following and more detailed investigations are needed to justify these suggestions:

(1) For the Zuider Zee of Netherland in Fig. 2.1, now the water quality is not very good, as the 125-km-long River IJssel sometime delivers low-quality water from the Rhine River to the CR. It is suggested that an intake structure is installed as the river's entrance (thick short yellow line [light grey line in printed version] in Fig. 2.6A); thus only high-quality water is allowed to enter the CR for storage, and any low-quality water will be discharged into the sea via the Rhine River.

(2) For Lakes Alexandrina and Albert of Australia shown in Fig. 2.3B and Table 2.2, the original design does not meet the requirements of

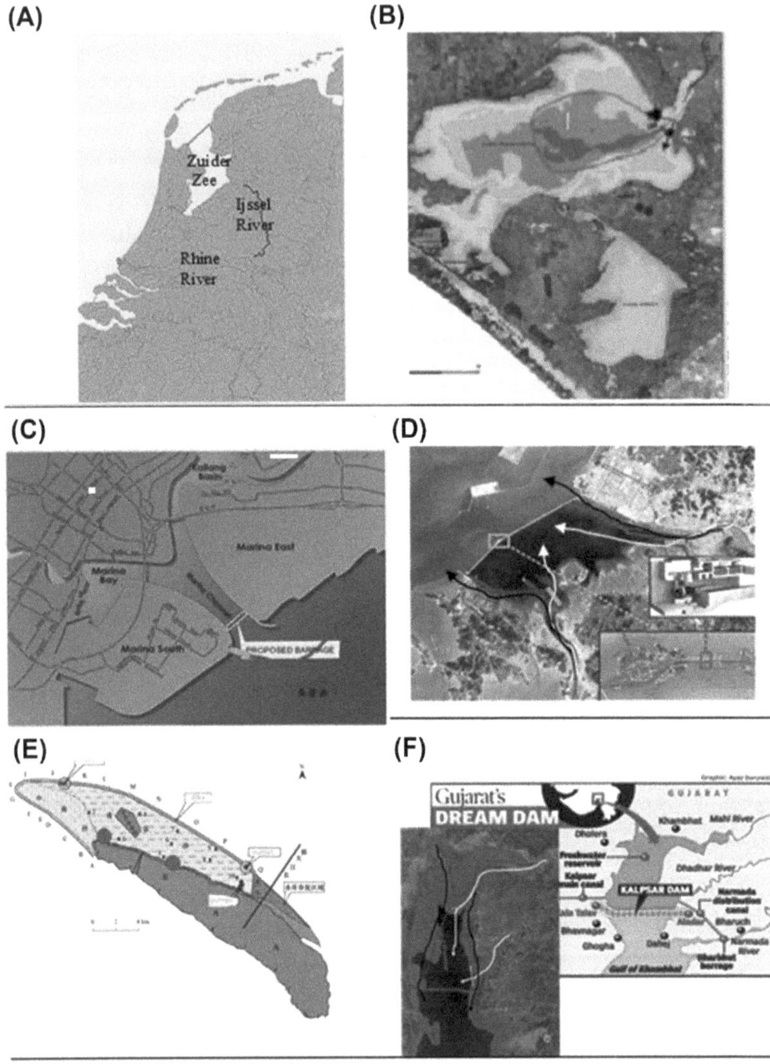

Figure 2.6 Modified existing/planning CRs' design where the dikes are represented by *red lines* (grey lines in printed version), hydraulic gates by *thick and short yellow lines* (light grey lines in printed version) (except (B) by *red short line* [grey short line in printed version]), good-quality water by *white lines* and poor-quality water by *black lines*. *CRs*, coastal reservoirs.

Eqs. (2.1) and 2.11, i.e. the storage capacity and surface area is too big and the CR's evaporation loss is too big, consequently, the salinity is very high as shown in Eq. (2.8) in extreme dry year. The modified design is shown in Fig. 2. 6B (Liu et al., 2018) where the dike is represented by a long red line (grey line in printed version) and the CR's intake is represented by the thick short red line (grey line in printed version), and the low-quality flow path is represented by black lines. Only deep water body is enclosed to satisfy Eq. (2.1). Liu's (2017) investigation shows that the design can improve Adelaide's water quality and also the lake's ecosystem.

(3) For the Marina Barrage of Singapore given in Table 2.2, the active storage volume is small even its surface area is big because the variable water depth is only 1–2 m to attract tourists. Consequently, every year huge amount of freshwater is dumped into the sea, but water is pumped back by a nearby desalination plant. To increase its storage capacity, it is suggested to store the dumped freshwater in the sea space as shown in Fig. 2.6C, i.e. the second-generation CR design should be applied in Singapore's Marina Barrage, i.e. the water body enclosed by the red (grey in printed version) curved line. If other CRs also use the sea space to store its excessive freshwater (Yang, 2003), Singapore should have sufficient water even without other sources such as seawater desalination, wastewater recycling (newater) and imported water.

(4) For the Sihwa Lake of South Korea as shown in Fig. 2.6D, two bypass channels are proposed (red lines: grey lines in printed version) to divert the polluted water (arrowed black lines) out of the reservoir, and the clean water (arrowed white lines) is allowed to enter the reservoir via the hydraulic gates (short and thick yellow lines [light grey lines in printed version]). If so, it is expected that the water quality in the reservoir will be improved significantly and the reservoir can be restored as a water supply reservoir.

(5) For Qingcaosha CR of Shanghai, China, in Fig. 2.6E, based on the optimal layout illustrated earlier, it is suggested that two more pumping stations (red dots: grey dots in printed version) will be helpful to minimize the possible stagnant water body and increase the water's mobility.

(6) Gujarat's CR, India, shown in Fig. 2.6F once completed will be the largest CR in the world. The underplanned Kalpasar Project envisages building a 30-km dam across the Gulf of Khambat in India for establishing a huge freshwater reservoir for irrigation, drinking and industrial purposes. It was identified as a promising site for tidal power generation by UNDP expert, Mr. Eric Wilson in the year 1975. Further studies have dropped this suggestion, but only for water resources development, and the dam length is reduced. By following the principle outlined in Eqs. (2.1)–(2.8), the shallow water body should be excluded from the CR, as shown in Fig. 2.6D. The exclusion of shallow water body is not only good for the water quality, but it is also a social problem. Saemanguem project in South Korea had negative social impacts because of too ambitious land reclamation plan. The same public opposition may happen in India if the reclamation is too greedy.

2.5 Can coastal reservoirs supply sufficient, high-quality and affordable water to the world with minimum environmental/social impacts?

Sufficiency: Different from other associations, IACRR's faith is that 'There is a sufficiency in the world for man's need' (quoted from Mahatma Gandhi). Table 2.4 shows that on average, every year the world loses 42,800 km^3/year of water into the oceans, and the agricultural, industrial and domestic total water consumption in 1990s was only 3240 km^3/year. Compared with the 1990s, in the 2010s, the annual withdrawals for agriculture, industry, municipalities and reservoir losses amount to 4430 km^3/year, of which roughly 54%, or 2285 km^3/year, is actually consumed, whereas the remaining 46% is returned after use like hydropower; thus the actual usage is only 5.3% of the water available.

Other organizations claim not enough water because they assume that reservoirs can only be built in a freshwater environment, not seawater environment, or they assume that run-off to the sea cannot be developed at river mouths and only desalination plants can do so. For example, FAO (1996) concluded that the upper limit of water appropriated for human use, including withdrawals and instream flow needs, totals 6780 km^3 per year. Obviously, from IACCR point of view, every drop of water can be developed, and the upper limit should be 42,800 km^3/year, but realistically

Table 2.4 Annual water availability and agricultural, industrial and domestic total water use in the 1990s.

Continent	Annual run-off (km^3) (1921 –1985 UNESCO)	Annual total water use (km^3)	%
Asia	13,508	1531	11%
Europe	2,902	359	12%
Africa	4,050	144	3.5%
North America	7,892	697	8.8%
South America	12,030	133	1.1%
Australia and Oceania	2,404	23	0.9%
World total	42,800	3240	7.4%

only a small amount of additional water is needed, say 1%—2% of the total run-off lost to the sea. It could be sufficient for future generations; then the water demand required to achieve SDG6 can be fully met.

Table 2.4 clearly shows that the world is not running out of water, but water is running out of our continents. CRs can supply sufficient water to coastal cities, which will attract more and more people to live; consequently, water in upstream reservoirs can be fully used by upstream community for their hydropower generation, irrigation or domestic purposes. This means that CRs can indirectly increase upstream water supply.

Affordability: Here, we compare the costs of four different water solutions in Australia, i.e. inland dam, desalination, wastewater reuse and CRs. The proposed A$1.6 billion Traveston Crossing Dam in Southeast Queensland, Australia, could supply 70 GL/year, and 76 km^2 of land is to be flooded permanently as the reservoir (Turner et al., 2007); its life span is about 100 years, during which span the total water supply is 7,000 GL; thus the capital cost is $0.23/kL (=A$1.6B/7B m^3). Its running cost is much cheaper compared with desalination or wastewater reuse.

Sydney desalination plant costed A$1.8 billion for its output of 90 GL/year. Its design life span is about 20 years; thus its capital cost is A$1.0/kL. This plant has been in 'standby mode' since its birth in 2010. This desalination plant needs the power of 4 kWh/kL (Stamatov and Stamatov, 2010); thus the energy cost is about $\$0.3 \times 4 =$ A$1.2/kL.

The western Corridor Project invested A$2.6 billion for water supply capacity of 130 GL/year; thus its capital cost is A$1.0/kL for 20-year design life span, required energy is less and A$0.6/kL is assumed. This gives that wastewater recycling method's energy cost is A$1.0/kL.

The Qingcaosha CR in Shanghai spent A$3.7 billion for the construction cost including 45-km-long dam, pumping system with the capacity of the pumps to supply a 200 m^3/s meeting an annual demand of 2600 GL, 114-km-long pipeline system, a 7.2-km underground tunnel (about 6 m in diameter) and two sluice gates with 70 and 20 m widths (http://news.hexun.com/2011-06-09/130379841.html). The designed life span of CRs is also 100 years; thus its capital cost is A$0.01/kL. Its energy cost is very low at A$0.05/kL.

In this study, inland dams' energy cost is assumed to be A$0.05/kL as same as the CR. Table 2.5 provides the comparison of costs among different methods, where the second row shows the name of projects. The total capital cost in row 4 means the cost at its beginning. Row 5 is its design capacity. Row 6, the capital cost per kL, is obtained by dividing the

Table 2.5 Cost comparison among different methods of water supply per kL of water.

	Inland dams	Desalination plant	Wastewater reuse	Coastal reservoir
Project discussed	Traveston Crossing Dam in QLD in 2007	Sydney Desalination Plant in 2010	West Corridor Project in QLD in 2008	Qingcaosha in Shanghai in 2010
Designed life span (year)	100	20	20	100
Total capital cost (A$ billion)	1.6	1.83	2.6	3.7
Water supply (GL/year)	70	90	130	2600
Capital cost (A$/kL)	0.23	1.0	1.0	0.01
Energy cost (A$/kL)	0.05	1.2	1.0	0.05

total capital cost in row 4 with the total water supply in its life span (i.e. the product of data in row 3 and 5). It can be seen that for 1 kL of water, the capital cost of desalination and wastewater recycling is about 100 times higher than the cost of CR; the energy cost is about 20 times higher.

Environmental impacts: The environmental impacts of different solutions, namely, desalination, inland dams and CRs are compared in this section. Three projects are chosen for comparison because they are earliest and have wide influence, i.e. the Hoover Dam and Zuider Zee, both were constructed in the 1930s. The desalination plant at Jeddah is the world's first large seawater reverse osmosis (RO) plant, which was commissioned in 1989.

When 'environmental impacts of Zuider Zee' were keyed in for topic search in the Web of Science on 1 January 2018, only one paper by Lammens et al. (2008) appeared. Their research reveals that the impacts of Zuider Zee reservoir include 'damming and fixing the water table prevented the development of emergent vegetation and caused steep water-land gradients'; other impacts are 'high nutrient loads, which cause phytoplankton blooms, the disappearance of aquatic macrophytes and intensive fishery'. However, when 'environmental impacts of Hoover Dam' and 'environmental impacts of Jeddah desalination plant' were keyed in the same database, the results were shown in row 5 of Table 2.6.

Similarly, other databases are used to search the article numbers, such as Scopus, google scholar and google; the results are present in rows 6 to 8. The 85-year Zuider Zee is long enough for researchers to infer CR's environmental impacts. Certainly, they attract far less concern than inland dams and desalination plants.

Social impact: If one compares the social impacts of inland dams such as Three Gorges Dam and CRs, the answer is very obvious as involuntary resettlement of communities is a big problem for inland dams. The resettlement of people generally is very costly; the complexity of resettlement and the restoration of livelihoods should not be underestimated. To build the Three Gorges Dam, China relocated 1.3 million people. Studies indicate that those directly affected were negatively impacted both socially and economically. This has the potential to cause conflict in villages where dispossessed people resettle and express opposing views to the community in which they resettle in. This can become a higher risk factor in multicultural countries where enclaves of people of dissimilar background settle together.

Local communities near CRs would benefit from access to a potable water source, being uplifted from the poverty line in particular circumstances, decreased threat of flood and secured livelihood for farmers. There would be limited, or no relocation of households and more jobs would be

Table 2.6 Article numbers in large databases by keying in 'environmental impacts of project'.

	Inland dams	Desalination plant	Coastal reservoir
Project	Hoover Dam	Jeddah desalination plant	Zuider Zee
Country	United States	Saudi Arabia	Netherlands
Competed year	1930s	1989	1930s
Web of Science	2	3	1
Scopus	8	4	1
Google Scholar	25,600	2,530	4,060
Google	443,123	41,300	35,700

created in the area during the development of the CR. Postcompletion of the CR, there is hope of increased health and well-being with increased recreational activities and a greater respect and appreciation for waterways. Business owners would benefit from increased tourism and economic standards in surrounding area.

The government should be required to create a sustainable approach and regulations regarding recreational and nonrecreational activities on the CRs. Additionally, they would benefit from a reduced flood water management scheme and funds supporting those impacted in the event that a flood does occur. CRs can be socially beneficial and have the potential to increase tourism within a region and thus hold economic value for communities. Designs with high aesthetic value such as Marina Barrage are popular for not only local community members but also tourists who visit the region. PUB Singapore's National Water Agency has encouraged ownership of the waterways and encouraged community involvement by allowing locals to use the space leisurely through water sports or surrounding walks.

An easy and simple way to compare social impacts of different solutions is to search Google imagine; when 'Marina Barrage' is keyed in, almost all pictures are happy, but other solutions have many negative imagines.

2.6 Conclusions

Rapid urbanization is growing worldwide, or more people migrate to cities, especially in coastal cities. The mega city Shenzhen, China, with about 12 million people could be a good example, which has grown from a small village of 175,000 in 1985. Consequently, it is predictable that most megacities will be situated near coastal areas, and more than 50% of people will live in these coastal cities, which will face water scarcity soon. It is certain that every coastal city will develop its run-off lost to the sea using CRs. It is good time to summarize the experience in CRs' design. This chapter concludes that the CR technology will dominate future water supply and it is a promising technology to quench the world's thirst if the design is good; the following conclusions support IACRR′ vision, i.e. CR can supply sufficient, high-quality and affordable water to the world with minimum environmental/social impacts:

(1) On average, the world just uses 5.3% of the annual available water resources. It is obvious that the world is not running out of water, but water is running out of our rivers. We have no water shortage problem, but the storage shortage problem. Seawater desalination plants are unnecessary in areas where the run-off is rich.

(2) A CR can collect not only all water from a catchment but also all contaminants from the catchment. For CR's high-quality water supply, it is important to be able to bypass unwanted/polluted water. If needed, wetland pretreatment should be considered to remove impurities.

(3) The CR's volume needs to match its targeted water demand; shallow water bodies should be excluded during site selection. Its inlet(s), outlet(s) and pumping stations' design should minimize stagnation within the water body.

(4) By comparing different water solutions, it is certain that the CR's energy cost and capital cost are far below than other solutions such as sea water desalination and wastewater reuse. CR technology also has minimum environmental/social impacts.

(5) It is suggested that all regions with abundant run-off to the sea consider the strategy of CRs in their planning stage and make a comparison with other alternatives such as new inland dams, desalination plants and wastewater reuse before a decision is made in practice.

References

Bae, Y.H., Kim, K.O., Choi, B.H., 2010. Lake Sihwa tidal power plant project. Ocean Engineering 37 (5–6), 454–463.

FAO, 1996. World food summit. In: Technical Background Document, Rome Italy, November, 13–17. http://www.fao.org/docrep/003/w2612e/w2612e07a.htm.

Gibbes, B., Quiggin, J., October 30, 2014. Dam hard: water storage is a historic headache for Australia. The Conversation. https://theconversation.com/dam-hard-water-storage-is-a-historic-headache-for-australia-33397.

Lammens, E., van Luijn, F., 2008. Towards ecological goals for the heavily modified lakes in the IJsselmeer area, The Netherlands. Hydrobiologia 599, 239–247. https://doi.org/10.1007/s10750-007-9202-2.

Lehner, B., Reidy Liermann, C., Carmen, R., Vörösmarty, C., Fekete, B., Crouzet, P., Petra Döll Endejan, M., Frenken, K., Magome, J., Nilsson, C., Robertson, J.C., Rödel, R., Sindorf, N., Wisser, D., 2011. High-resolution mapping of the world's reservoirs and dams for sustainable river-flow management. Frontiers in Ecology and the Environment 9 (9), 494–502. https://doi.org/10.1890/100125.

Liu, J., Yang, S.-Q., Jiang, C., 2015. Coastal reservoirs strategy for water resource development – a review of future trend. Journal of Water Resource and Protection 5 (3A), 336–342.

Liu, J.L., 2017. Hydrodynamic and Salinity Simulation in the Lower Lakes, South Australia and Proposed Coastal Reservoir (Ph.D. thesis). Univ. of Wollongong, Australia.

Liu, J., Sivakumar, M., Yang, S.Q., Jones, B.G., 2018. Salinity modeling and management of the lower lakes of the Murray-Darling basin. WIT Transactions on Ecology and the Environment 228, 257–268.

Stamatov, V., Stamatov, A., 2010. Long-term impact of water desalination plants on the energy and carbon dioxide balance of Victoria, Australia: a case study from Wonthaggi. Water and Environment Journal 24, 208–214.

Turner, A., Hausler, G., Carrard, N., Kazaglis, A., White, S., Hughes, A., Johnson, T., 2007. Review of Water Supply-Demand Options for South East Queensland. Institute for Sustainable Futures, Sydney and Cardno, Brisbane.

Wang, J., Yang, S.Q., Liu, P.W., 2009. Preliminary research on the programme in comprehensive governancing of Dongting lake. Environmental Science and Technology 32 (11), 76–79 (in Chinese).

Yang, S.Q., 2003. Potential water resources in Singapore. Journal of Water Supply: Research and Technology 52 (6), 425–434.

Yang, S.-Q., 2004. In: Global and China's water crisis and its solutions in the 21st century. Tianjin University Press, Tianjin, ISBN 7-5618-2069-0 in Chinese.

Yang, S.-Q., 2015a. A preliminary feasibility study for a backup water supply coastal reservoir in Southeast Queensland, Australia. Research and Technology – AQUA 64 (4), 470–485.

Yang, S.Q., 2015b. Coastal reservoir-a technology that may dominate future water supply. J. Water Resource and hydraulic Eng. 4 (4), 388–397.

Yang, S.Q., 2016. Technology of coastal reservoir and its contribution to the world water crisis. Proc. of International Conference on Water, Environment. Environment, Energy and Society (ICWEES-2016). March 15–18, Bhopal, India.

Yang, S.Q., Liu, P.W., 2010. Strategy of pollution prevention in Taihu lake and its effects analysis. Journal of Great Lakes Research 36 (1), 150–158.

Yang, S.-Q., Zhang, Z., Ng, B.C., Ong, B.C., 2001. Coastal Reservoirs. Singapore Patent No. 200504653-200504657.

Yang, S.-Q., Lin, P.-Z., 2011. Coastal reservoir by soft-Dam and its possible applications. Recent Patents on Engineering 5, 45–56.

Yang, S.-Q., 2017. Will coastal reservoirs dominate future Australian water supplies? Journal of Sustainable Urbanization, Planning and Progress 2 (2), 4–15.

CHAPTER 3

Water quality considerations: from catchment to coastal reservoir

Muttucumaru Sivakumar[1], Brian G. Jones[2], Shu-Qing Yang[1]
[1]School of Civil, Mining and Environmental Engineering, University of Wollongong, Wollongong, NSW, Australia; [2]School of Earth, Atmospheric and Life Sciences, University of Wollongong, Wollongong, NSW, Australia

3.1 Introduction

Water is indispensable to all life forms. We must have adequate water for all our essential needs, which must be of suitable quality for the intended purpose. Increase in production and consumption patterns, land use changes, urbanization, industrial and agricultural practices, as well as impact of climate change, are increasingly affecting the quantity and quality of available water. The United Nations Sustainable Development Goal 6 (SDG6) calls for sustainable management of water and sanitation for all by 2030. To ensure that the SDG6 goals can be achieved, we need to look for innovative approaches, and it is proposed that coastal reservoirs (CRs) are a potential solution that can provide not only adequate quantity of water but also water of good quality.

A CR is defined as any structure designed to capture fresh river flow before it enters the sea and mixes with salt water. These water retention structures, by the very nature of their design, have to be placed at the tail end of the catchment; hence they will be naturally affected by the various hydrologic, hydraulic, hydrodynamic, tidal and other environmental processes that occur upstream of a CR. Hence, it is vital that a deep understanding of these processes is essential so that CRs can be designed, constructed and operated in a sustainable manner. A number of simple water retention infrastructures have been built in the past to retain fresh nonsaline river water before entering the sea. These include barrages, weirs, bunds and vented dams and mostly have some form of gate operation to control flow (Fig. 3.1). This method of storing water inline with the river flow would be considered as a 'first-generation' CR. The first-generation CRs are particularly susceptible for catchment and water quality

Sustainable Water Resource Development Using Coastal Reservoirs
ISBN 978-0-12-818002-0
https://doi.org/10.1016/B978-0-12-818002-0.00003-4

Figure 3.1 Types of first-generation coastal reservoirs: (A) barrage, (B) weir, (C) bund, and (D) vented dam.

deterioration as they are at the mercy of the water quality processes that occur upstream. A good example of the earliest known CR is Lake Alexandrina in Australia at the tail end of the Murray-Darling River system. This was originally a natural lake prior to 1930s, and the construction of five barrages turned it into a freshwater first-generation CR. Since then, the water quality of Lake Alexandrina has been affected significantly by the upstream river and catchment processes such as land use changes, irrigation returns, floods, droughts and water diversion schemes.

When nonsaline freshwater can be stored offline from the river, such as in artificial channels or ponds to recharge groundwater or in a number of small to very large reservoirs (e.g. Qingcaosha Reservoir in the Yangtze River estuary near Shanghai), they are classified as 'second-generation' CRs. These reservoirs, as shown in Fig. 3.2, have the advantage of capturing river water only when the water is suitable for the intended purpose. Second-generation CRs can be potentially designed for multiple purposes such as water supplies for domestic, agricultural or industrial use; upstream flood reduction to minimize inundation of urban areas; urban regeneration and water front land development to enhance high-quality housing and water-based leisure activities; pumped hydro storage to generate renewable energy during peak periods and others. In each of these cases, the water quality requirements would be different.

Figure 3.2 Map showing the locations of four-second generation coastal reservoirs in the Yangtze River estuary, Shanghai. *(Source: Google Earth, 2019.)*

The main aim of this chapter to guide designers into the various contaminants that can potentially affect the water quality inflow into a CR. A brief review of water quality processes that originates in a given catchment and ends up in a CR is undertaken. Recommendations will be made on the importance of water quality and how it should be considered in the selection of sites, design and operation of a CR.

3.2 Contaminants affecting water quality and catchment processes

The water quality at the intake point of a CR can be affected by a number of contaminants that may be present in the flood water. These may include salts, sediments, nutrients, microorganisms, organic and inorganic chemical toxicants including heavy metals, biodegradable organic matter and emerging micropollutants such as microplastics, pharmaceutically active compounds and endocrine disruptor chemicals. The concentration of each contaminant as measured by well-established water quality parameters will vary depending on the generation, transport and other physical, chemical and biochemical transformation kinetics of a given catchment river system. Potential sources and desired concentrations of the some of the key contaminants are described here.

3.2.1 Salts

In the freshwater environment, salt may be potentially added to the water by a number of anthropogenic activities from point sources (domestic and industrial effluents, leachates, etc.) and from nonpoint sources such as urban runoff, land clearing and agricultural activities including irrigation returns. The level of salt or salinity in a water body can be measured by the amount of total dissolved solids (TDS) present in the water measured in mg/L. While TDS requires gravimetric determination in a laboratory, it can be indirectly measured continuously by an electrical conductivity (EC) measuring probe in μS/cm. Depending on the water body's ionic composition, a direct correlation can be established between TDS and EC. For example, McNeil and Cox (2000) analyzed a large number of surface water samples in streams and rivers (around 34,000 samples) and found that total dissolved ions in water can be calculated by multiplying conductivity values by a factor between 0.63 and 0.72 depending on the conductivity range between 10 and 1000 μS/cm. TDS is approximately equal to $\frac{2}{3}\cdot$ EC has been used in the literature. In terms of a guideline value for water supply purposes, it would be good to limit the conductivity to less than 1000 μS/cm. Many river or lake systems around the world that monitor water quality measures EC (https://riverdata.mdba.gov.au/peechelba), and these data can be made available in real time. Flood events while having low salinity values still can bring significant amount of salt flux. During low flow periods, engineered systems such as salt interception schemes play a key role in diverting salt away from river systems such as in Murray River in Australia (MDBA, 2013).

From designing and selecting a suitable location of a CR for water supply perspective, one of the biggest problems is avoiding the saltwater intrusion coming from the sea or an estuary. This would depend on the flow patterns of the river and the tidal hydrodynamics of the estuarine system. A case study of this problem between river discharge and saltwater intrusion is well illustrated by Xu et al. (2018). For the Yangtze River estuary, several second-generation CRs exist (Fig. 3.2). Long-term monitoring of EC near potential water intake locations may be required to determine the number of days to avoid when EC exceeds the acceptable values as a suitable water supply source. During these periods, raw water should not be drawn into the CR.

Depending on the location where a CR is sited, during the early stages of the filling, saltwater may have to be removed over a period of time. This can potentially take up to a few years depending on the water exchange processes. In addition, saltwater can enter the CR via subsurface flow as well as potential overtopping during extreme storm events. It is possible to minimize or overcome these problems by suitable engineering measures.

3.2.2 Suspended sediments

A number of natural and anthropogenic activities generate sediments in catchments and river systems. Sediments in suspension can be carried by streams over large distances. Floods and especially large storms have the potential to transport significant amounts of both bed and suspended sediment load. Fine sediments or suspended particulate matter can also carry nutrients that can cause potential eutrophication, and other harmful substances are persistent organic chemicals, heavy metals and other emerging pollutants such as microplastics. Sediments suspended in water give rise to turbidity, prevent light penetration, will settle to the bottom of the water body and also increase the cost of potential downstream water treatment.

In environmental engineering practice, suspended sediment is measured by the parameter total suspended solids (TSS). TSS is a measure of the amount of suspended solids in water. It is measured in a laboratory after taking a representative grab sample in the field or a composite flow-weighted field sample in some cases. In the laboratory analysis, a representative sample volume is taken from the field sample and filtered through a 2-μm filter paper. All solids remaining on the filter paper after a suitable drying process (Rice et al., 2017) are taken as the TSS. While TSS has been widely used by various water authorities as a measure of the suspended sediments, it can give rise to erroneous results. In particular when the suspension has particles greater than 62.5 μm (fine sands), TSS can underestimate the actual concentration of suspended sediments.

In river engineering practice, however, the suspended sediment concentration (SSC) is measured in the field using an isokinetic sampler (ASTM, 2007) that provides a much more accurate sampling method based on flow and width of the open channel since sampling is done at multiple points across the channel. When water samples are taken to the laboratory, suspended sediment concentration is determined with the whole sample, and in some cases, a particle size distribution is also undertaken. Hence, SSC should be a much more representative parameter of the actual sediment levels than TSS in rivers and streams (Ellison et al., 2014; Groten and Johnson, 2018).

Turbidity is a water quality parameter that is widely used to indicate water clarity. It is a measure of the ability of the water to scatter light and measured in Nephelometric Turbidity Units (NTU) that can be calibrated with a stable Formazin polymer. The turbidity measurement is affected by particle size, shape and concentration and can be instrument dependent.

However, for a given water system, turbidity can be used as a surrogate to measure SSC, or TSS as suitable correlations can be obtained between turbidity and SSC and/or TSS. Hence, SSC or TSS can be indirectly measured in rivers in real time using a suitable turbidity probe provided the probe can be maintained to minimize fouling.

From the river ecology point of view, environmental quality standards for suspended solids can be set for rivers used for water supply purposes. For example, Japan Ministry of Environment (MOE, 2019) has set a daily average value of TSS $\leq$ 25 mg/L for water supply classes 1, 2 and 3. Depending on the water supply classes, the types of unit water treatment processes required will vary to bring the raw water to drinking water quality standard. According to the Japanese standards, water supply class 1 requires a minimum of simple filtration, water supply class 2 requires sedimentation and filtration and water supply class 3 requires pretreatment and other advanced methods. It is a well-established practice that all drinking water treatment processes will also undergo some form of disinfection before water is supplied to the community.

3.2.3 Nutrients

Nutrients such as nitrogen and phosphorus play a key role in productivity of water systems. Excess amount of nutrients can give rise to harmful algal blooms (HABs), eutrophication of the water body that can cause loss of biodiversity, acidification, fish mortality, as well as affect the oxygen content leading to hypoxia. The human-induced sources of nutrients can originate from various catchment land uses, such as sewage effluents, agricultural fertilizers, effluents from aquaculture, fossil fuel combustion, animal manure or wastes, erodible soils, industrial wastewater, landfill leachate and detergents. Other natural sources include atmospheric deposition and decaying vegetation. Unlike salts and sediments that are considered conservative pollutants, nutrients are nonconservative pollutants. Nutrients undergo transformation based on various biochemical processes that occur in the water body. For biological activity, both the concentration and flux or load (flow $\times$ concentration) of nutrients will be important. During high river flows, the nutrient concentrations may be low, but the nutrient load can be very high.

Nutrients are of two kinds, namely, nitrogen and phosphorous. The different forms of nitrogen (total nitrogen [TN]) that are present in the water include organic nitrogen (Org-N), ammonium nitrogen (NH_4−N), ammonia nitrogen (NH_3—N), nitrite nitrogen (NO_2—N) and nitrate (NO_3—N) nitrogen. The first three terms added together are called total Kjeldahl nitrogen (TKN), whereas the latter four terms added together are called total inorganic nitrogen (TIN). In addition, TN can be present in dissolved form, called total dissolved nitrogen (TDN), and the suspended form called the total particulate nitrogen (TPN). The various nitrogen forms can undergo transformation; first, organic nitrogen is converted to ammonium ions (ammonification), and subsequently, ammonium is oxidized to nitrite and nitrate via a process called nitrification, and the final stable form of nitrogen is nitrate nitrogen. Under suitable environmental conditions, nitrate nitrogen can be reduced to nitrogen gas by the denitrification process. Each of the individual nitrogen parameters at elevated concentration can give rise to toxicity to aquatic life and/or health effects on humans. For example, to prevent illness in babies, the so-called blue baby syndrome known as *methemoglobinemia*, US national water quality criteria recommend nitrate concentrations not to exceed 10 mg/L (USEPA, 2017), whereas for the protection of ecosystem health, the Japanese environmental standard restricts nitrate concentration to $\leq$ 0.2 mg/L (MOE, 2019).

Phosphorous (total phosphorous [TP]) is also present as both organic or inorganic forms depending on their origin. The soluble form of inorganic phosphorus is referred to as the soluble-reactive phosphorous, which is readily bioavailable and becomes one of the important parameters in plant growth. Particulate form of phosphorous can be bound to sediments, and it can be potentially recycled between the water column and sediment. A good review of the various physicochemical processes involved in the phosphorous cycling in rivers and the implications of catchment management practices are found in Withers and Jarvie (2008). With reference to providing a numeric standard value for phosphorous in streams, many countries are moving towards developing site-specific values. In the United Kingdom, soluble-reactive phosphorous values for rivers are recommended to be below 30—50 μg/L. While in Australia (ANZG, 2018) and the United States, site-specific trigger values are obtained since the ecosystem types are different in each region, and background water quality can vary widely.

3.2.4 Other conventional pollutants

Conventional pollutants that are emanating from both point sources (e.g., sewage effluents, industrial discharges, etc.) and nonpoint sources (such as agricultural runoff, urban stormwater runoff, etc.) affect the water quality of a river. These pollutants usually can be measured by the water quality parameters such as temperature, pH, biochemical oxygen demand (BOD_5) or chemical oxygen demand (COD), total organic carbon (TOC), coliform bacteria of fecal origin and dissolved oxygen (DO) concentration. For a given body of water, it is possible to obtain a correlation between BOD (takes 5 days to measure) and COD (about 2 h). Both BOD and COD measurements require reagents. TOC can be measured much more accurately than BOD_5 and COD. Recent studies (e.g., Lee et al., 2016) indicate that TOC can be potentially considered as a surrogate to COD than BOD_5 particularly in rivers. DO in rivers is affected by various sources (reaeration, rainfall, oxygen saturated branch channel flows, photosynthetic activity) and sinks (biodegradable organics from sewage, urban runoff and industrial effluents as well as agricultural sources including piggery wastes). DO is an important water quality parameter for aquatic flora and fauna that include fish and invertebrates. For water supply purposes, the recommended DO values in the river should be above 7.5 mg/L (MOE, 2019).

Increasingly non—reagent-based UV-VIS-based spectral sensors are coming to the market where online real-time measurements of selected water quality parameters are becoming a reality. It is now possible to measure multiple water quality parameters for rivers, for example, nitrate, nitrite, COD, DO, TSS, etc., by these spectral and fluorescence methods (see xylemanalytics.com), which can be very useful as an early detection of potential poor water quality that can be bypassed from the CR.

3.2.5 Emerging pollutants

There are a multitude of emerging pollutants that originate from the upstream catchment sources and can potentially enter a CR using a variety of pathways. These pollutants include animal drugs (ADs), endocrine disruptors (EDs), engineered nanoparticles (ENPs), fire retardants (FPs), illicit drugs (IPs), microplastics (MPs), personal care products (PCPs), pesticides and herbicides (PHs) and pharmaceutically active compounds (PhACs). A summary of the emerging pollutants sources, key contaminants and their potential impact to a CR is summarized in Table 3.1 based on recently published literature. Almost all of these contaminants are found at very low concentrations in surface water, and it is expected that during flood events in a river system, these contaminants would be at even lower levels due to the effect of dilution. While some of these contaminants are

Table 3.1 Emerging contaminants that affect surface water quality.

Contaminant category	Source	Selected emerging contaminants	Refs.	Potential contamination and impact on CR
Animal drugs	Widespread use in livestock industry, excreted by animals and can enter via soil and groundwater to surface water	Sulfamethoxazole, sulfamethazine, lincomycin, metronidazole, trimethoprim, sulphonamides, macrolides, carbamazepine, fluoroquinolones, sulphonamides	Snow et al. (2019)	Highly unlikely unless CR is situated very close to intense animal farms.
Endocrine disruptors	Food packaging, plastics, pesticides, cleaning agents, phthalates	Phenolic compounds such as alkylphenol, bisphenol-A, chlorophenols, nitrophenol and nonylphenol; 17b-estradiol, diethylstilboestrol, ethinylestradiol, triclosan	Casatta et al. (2016), Tijani et al. (2016)	Detrimental effect on aquatic species. CR is unlikely to be affected, as the concentration in floodwater would be negligible.
Engineered nanoparticles	Consumer products (socks), cosmetics, fuel additives, PV cells, sunscreens	Carbon nanotubes, CeOx, silver, fullerenes, SiO_2, TiO_2, ZnO	Troester et al. (2016)	Found in very low concentration depending on the production levels in the catchment. May aggregate and accumulate in the sediments. Effect on CR water is highly unlikely.

Continued

Table 3.1 Emerging contaminants that affect surface water quality.—cont'd

Contaminant category	Source	Selected emerging contaminants	Refs.	Potential contamination and impact on CR
Fire retardants	Building and construction materials, electronics, foams, furniture	Organophosphorus flame retardants	Pantelaki and Voutsa (2019)	Higher concentration detected in streams near industrial plants. Must avoid CR location nearby chemical industrial activities.
Illicit drugs	Illegal use of narcotics	Amphetamines, cannabinoids, cocaine, opioids and their metabolites	Yadav et al. (2017)	Found in very low concentration. Advanced treatment processes can remove most of these contaminants. CR water is unlikely to be affected.
Microplastics	Industrial products, plastic wastes, synthetic textiles	Polyamide, polyethylene, polyethylene terephthalate, polypropylene, polystyrene	Akdogan and Guven (2019), Koelmans et al. (2019)	Likely to be present in both sediment and water. Good catchment management and CR water treatment may be required.
Personal care products	Cleaning detergents and disinfectants, conditioners, cosmetics, insect repellents, perfumes, shampoos, soaps, toothpastes	Bisphenol-A, methylparaben, propylparaben, salicylic acid, triclosan	Peña-Guzmán et al. (2019)	Enter surface water via wastewater treatment plant discharges. Detrimental effects on aquatic organisms when concentrations exceed toxicity level. Cumulative effects on human are unknown.

Pesticides and herbicides	Agriculture, forestry, horticulture, urban runoff	Atrazine, bromacil, diethylamine, dimethoate, metalaxyl, methomyl, trichlorfon	López-Doval et al. (2017)	Drinking water guidelines exist to most pesticides (ANZECC).
Pharmaceutically active compounds	Antibiotics, antidepressants, antiinflammatory and analgesics, antihistamines, β-blockers	Caffeine, ciprofloxacins, carbamazepine, diclofenac, ibuprofen, ketoprofen, ranitidine	Tijani et al. (2016) Peña-Guzmán et al. (2019)	Mostly found in urban water. Combined effects in human are unknown. CR water may likely to have at trace concentrations depending on the catchment.

CR, coastal reservoir.

known to cause damage to aquatic species, the effect on human health at very low concentrations, including their cumulative effect, is largely unknown. Hence, it is possible to decipher that their effect on a CR's intake water quality is potentially negligible. However, sustainable and integrated catchment management practices are essential to control and potentially eliminate their sources to surface waters if they are used for drinking water supply. If needed, suitable but costly advanced water treatment unit processes can be adopted to remove the emerging contaminants to ensure that water quality for the intended use is assured.

Selected water quality parameters and their guidelines for river waters used for water supply purposes around various countries are compiled and given in Table 3.2. For the purposes of environmental protection as well as for human health preservation, countries such as Australia, New Zealand and the United States have developed water quality frameworks and suitable methodologies to establish site-specific guideline values for the various contaminants. Japan, the European Union and the United Kingdom have developed environmental standards for ecosystem protection. From the CR intake water quality point of view, acceptable guideline values of the various contaminants such as salts, suspended sediments, nutrients, other conventional pollutants and toxic organics and inorganics are given in Table 3.2. In considering water flow in a river, they are essentially characterized by the water velocity, water height and volumetric flow rate. However, when water quality is considered, this is much more complex, and over a minimum of 150 contaminants need to be considered when it comes to water sources to be used for human consumption purposes.

For design of CRs, quantity of water available is critical; however, the spatial and temporal variation of water quality will play the most important role in site selection, extraction, gate operation, size and shape, etc. Hence, real-time water quality monitoring of key water quality parameters may be of paramount importance.

3.3 Water quality processes in catchment and river channels

3.3.1 Catchment land use and water quality

In natural and forested catchments, the surface water emanating will have excellent water quality free of contaminants. The bushes, trees, canopy and the leaves present in the forest floor create an ecosystem that creates less

Table 3.2 Typical water quality guidelines of selected parameters for river waters used for water supply purposes.

Parameter	Unit	Australia[1]/New Zealand (ANZECC, 2000; ANZG, 2018)	Japan[2] (MOE, 2019)	UK[3] (DEFRA, 2014)	USA[4] (USEPA, 2017)	Remarks
Biochemical oxygen demand	mg/L	None specified	≤ 3	5	None specified	[1]Southeast Australia, trigger values
Dissolved oxygen	mg/L	85%–110% sat	≥ 7.5	>5	Site-specific value to be determined	[2]Water supply classes 1 and 2
Total nitrogen	μg/L	500	≤ 200	None		[3]High- to good-quality river, 90th percentile standard
NH_4–N	μg/L	20		600[6]		
NO_3–N	μg/L	40[5]			10,000	
Total phosphorous	μg/L	50	None specified		State/location specific	[4]Protection of human health
Soluble-reactive phosphorous	μg/L	20		30–50		
pH	–	6.5–7.5	6.5–8.5	6–9	5–9	[5]Oxides of nitrogen
Total dissolved solids (salinity)	mg/L	125–2200	None specified	1000	250	[6]Total ammonia
Conductivity	μS/cm	20–250				
Total suspended solids	mg/L	None specified	≤ 25	≤ 25	None specified	[7]Additional 21 substances are being monitored
Turbidity	NTU	6–50			1–10	

Continued

Table 3.2 Typical water quality guidelines of selected parameters for river waters used for water supply purposes.—cont'd

Parameter	Unit	Australia[1]/New Zealand (ANZECC, 2000; ANZG, 2018)	Japan[2] (MOE, 2019)	UK[3] (DEFRA, 2014)	USA[4] (USEPA, 2017)	Remarks
Total coliform	MPN/100 mL	None specified	≤50 −1000	None specified	None specified	[8]Additional 6 substances are being monitored
Toxic organics	μg/L	Default guideline values are available for 135 organic and inorganic toxicants (www.waterquality.gov.au)	Standards available for 15 substances[7]	Proposed standards available for 23 priority substances	Recommended criteria for 146 substance	[9]Site-specific trigger values need to be developed
Toxic inorganics	μg/L		Standards available for 9 substances[8]	Proposed standards available for 6 priority substances		[10]No standards or guidelines are available to date
Emerging contaminants other than PHs (Table 3.1)	—	N/A[10]				

PHs, pesticides and herbicides.

water runoff, very little erosion and hence less turbidity, which significantly minimizes subsequent water treatment costs. In a survey undertaken by the American Water Works Association to determine the effect of catchment forest cover on treatment cost, it has been found that a 1% increase of forest cover reduces the turbidity by 3% (Warziniack et al., 2017). Forest cover can also be reduced by natural events such as bush fires. In a review paper, Smith and coworkers found that after bush fires in forested catchments used for water supply, water quality and in particular, sediments as well as pollutants associated with sediments such as nutrients, trace elements and organic carbon can be significantly affected (Smith et al., 2011). When catchment undergoes land clearing for agriculture and urban development, including industrial and commercial purposes, water quality of the adjacent water courses can be potentially affected as depicted in Fig. 3.3.

The effect of land use changes on water quality can be best illustrated in catchments that undergo the urbanization process. Many statistical techniques can be used to model this process. The simplest model is the constant concentration model, which averages event mean concentration (EMC) data from catchments within a homogeneous domain to form a single representative estimate of the pollutant concentration.

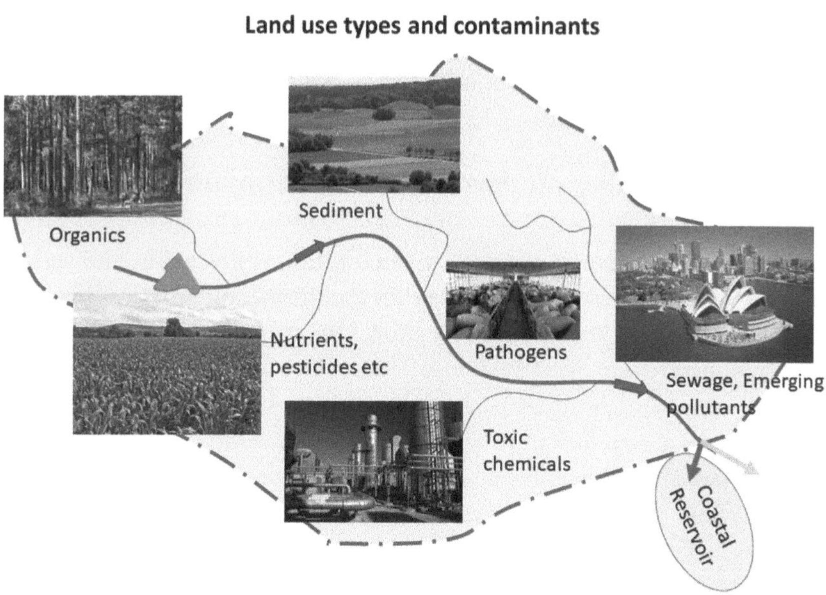

Figure 3.3 Catchment land use types and potential contaminant sources.

EMCs can be calculated using flow-weighted mean concentrations as defined in Eq. (3.1):

$$\mathrm{EMC} = \frac{\sum_{i}^{n} C_i Q_i}{\sum_{i}^{n} Q_i} \tag{3.1}$$

where C_i is the concentration of sample i, Q_i is the volumetric flowrate of sample i and n is the total number of samples for a given storm. Using the largest water quality data set available from US Nationwide Urban Runoff Program (NURP), May and Sivakumar (2009a) have analyzed over 1000 storms and found that EMC of a given pollutant can be related to different land uses. Fig. 3.4 shows that EMC of four different pollutants concentrations (dissolved solids, suspended solids, total nitrogen and total phosphorous) emanating from land use industrial (LUI), land use agricultural (LUAGR), mixed land use (LUM), land use nonurban (LUN), land use industrial and commercial (LUIC) and land use residential (LUR). It is clear that the industrial land use had the highest EMCs of all four contaminants studied followed by agricultural land use. When it comes to heavy metals transport from urban catchments, they occur largely from impervious areas (May and Sivakumar, 2009b). It is possible to show that for predicting urban water quality of unmonitored catchments, multiple regression models are more applicable than even models based on artificial neural network (ANN) (May and Sivakumar, 2008).

3.3.2 Water quality models for coastal reservoir intake water

Water quality of a river system at a given location is affected by the various inputs from catchments and the in-stream transformation processes that take place upstream of the chosen location. In the absence of long-term measured data, one must resort to modelling the water quality using physically based or lumped parameter models. A catchment-wide water quality model is required to predict the likely pollutant concentrations and load that are likely to enter a CR. A free-to-access popular model like SWAT—Soil and Water Assessment Tool (https://swat.tamu.edu/)—although originally developed for agricultural catchments has the ability to predict flow, sediment, salt and nutrient concentrations and contaminant load at an intake point of a proposed CR. For predominantly urban catchments, another free-to-access model HSPF (Hydrological Simulation

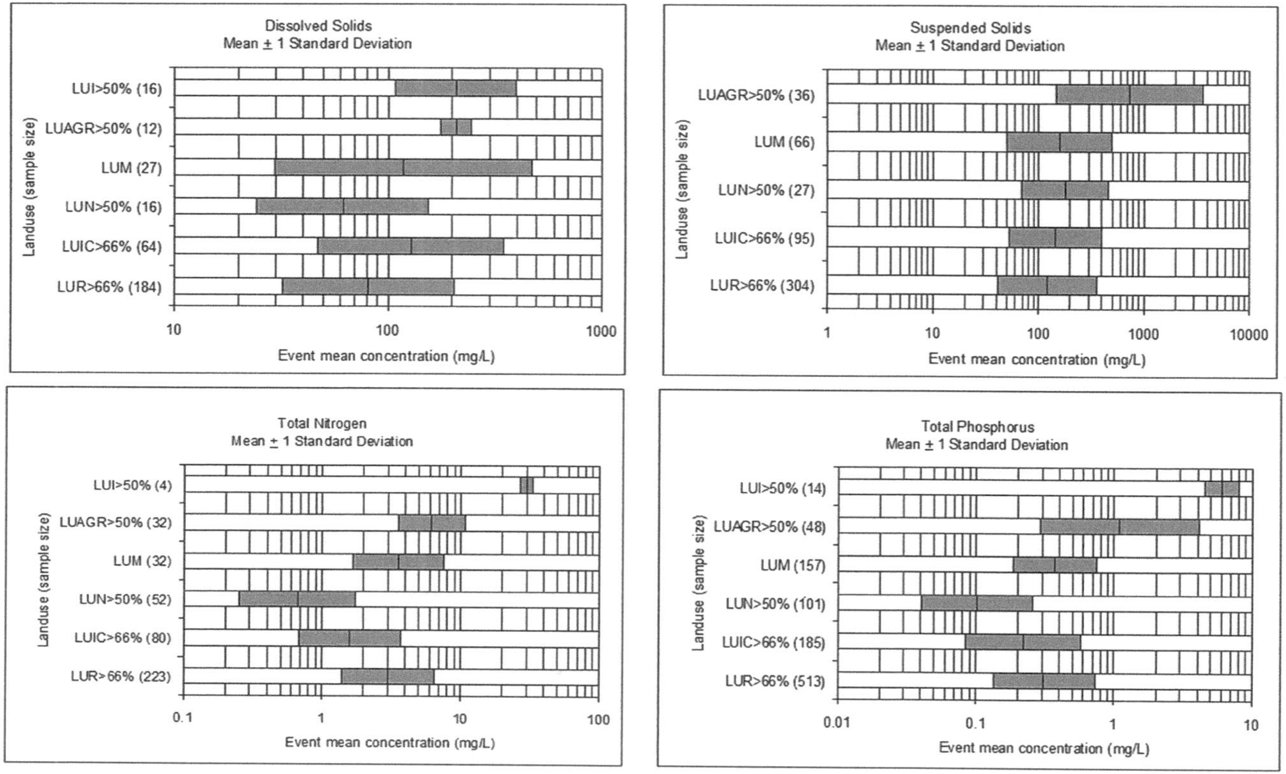

Figure 3.4 Land use and EMC of dissolved solids, suspended solids, total nitrogen and total phosphorus in urban catchments. *EMC*, event mean concentration.

Program Fortran) can be readily used for water quality prediction. Both of these models have been developed in the United States and applied both in the United States and other countries. Comparing the predictive performances of both models has produced some mixed results. For a 12,048 ha catchment size, the prediction of flows and suspended sediment load yield similar results from both models, but HSPF predicted marginally better when a time step of more than a month is considered (Im et al., 2007). However, the prediction of faecal coliform concentrations for a 1560 ha catchment has shown that SWAT provides a more accurate estimate, whereas the HSPF model has produced more accurate daily flows over a 7-year study period (Chin et al., 2009). A more complex model like INCA, developed in the United States, not only can predict catchment scale water quality and nutrient loads load but also can incorporate the impact of climate change and other socioeconomic factors. Such studies have been successfully undertaken for major rivers such as the Mekong River basin (Whitehead et al., 2019). A good review of the various catchment-wide water quality models available in literature, including the model eWater Source developed in Australia, has been recently summarized by Fu et al. (2019).

3.4 Water quality processes in lakes and reservoirs

A CR is a contained body of water. The water quality within the reservoir will be affected by the inflows and outflows and their contaminants including mass flux, residence time of the water in the reservoir, shape and size of the reservoir including reservoir bathymetry, reservoir bottom sediment–water interaction, surface water–groundwater interaction, meteorological and hydrodynamic conditions within the reservoir and potentially saltwater spray during extreme storm events.

Liu et al. (2018) have modelled the hydrodynamics and water quality (salinity) of the Lower Lakes in South Australia. Lake Albert and Lake Alexandrina are often referred to as the Lower Lakes and are located at the terminus of Murray River. Lake Albert is a terminal lake connected to Lake Alexandrina by a narrow channel. Lake Alexandrina is a broad and shallow (mean depth 2.86 m, maximum depth 4.75 m), well-mixed, regulated water body, with a surface area of approximately 650 km^2 and volume of approximately 1620 GL at + 0.7–0.75 m AHD (Australia Height Datum). Five barrages separate Lake Alexandrina from the Coorong Lagoon and the Murray estuary, which are the Goolwa, Mundoo, Boundary Creek,

Ewe Island and Tauwitchere barrages. These barrages were constructed in the 1930s, and the Lower lakes are essentially considered as a first-generation CR.

There were five forcing mechanisms within the Lower lakes (wind, river inflow, solar heating and cooling, barometric pressure and gravity). Wind and river inflow are the most influential mechanisms in shallow lakes. Wind forcing is a key factor determining lake circulation and a major energy source for horizontal motion and vertical mixing (Ambrosetti et al., 2003). When the wind blows over a lake, it exerts a surface shear stress on the water surface, results in momentum transfer from the air into the water and causes the surface water to move in the direction of the wind. Surface waves transport and dissipate a portion of wind energy, whereas the remaining energy forms large-scale currents, with typical surface water speeds of about 1.5%–3% of the wind velocity (Wuest and Lorke, 2003). In shallow lakes, wind-induced turbulence may occur at all depths and, therefore, can significantly enhance nutrient entrainment from the sediment bed.

Flushing time reflects the time elapsed for dissolved substances to be transported from one point to another, making it a useful time scale for describing the complex hydrodynamic and biogeochemical processes in large shallow lakes (Gong et al., 2009). Flushing process also could be used as an indicator of flow spatial pattern distribution. Usually, the shorter flushing time at a given lake region means the better self-purification with the lake region.

The flushing time for the Lower Lakes was modelled (Liu, 2017) using the Hydrodynamic and Transport Modules of the MIKE 21 Flow Model FM. The hydrodynamic (HD) module provided the hydrodynamic basis for the computations in response to a variety of forcing functions in the lakes. The transport module can simulate the spreading and fate of dissolved or suspended substances in an aquatic environment under the influence of the fluid transport and associated dispersion processes, which are typically applied in flushing studies, tracer simulations and water quality studies like salinity. The two-dimensional hydrodynamic module was set up initially with an assumed level of concentration (here the initial concentration was given as 100) in the whole domain of the model area, and the concentration for inflow was given as 0 to examine the flushing ability. The background for the model with reference to river flow was that there was a period of severe drought in the Murray Darling Basin from 2003 to 2009. High rainfall appeared through 2010 and early 2011, which resulted in significant flows in both the Darling and Murray River systems for the first time in the past 10 years. These high flows refilled the Lower Lakes and flushed considerable amounts of salt from Lake Alexandrina.

Fig. 3.5 shows that when the Murray River flows into Lake Alexandrina, the area from Murray River entrance to 4 km west of Pomanda was completely flushed within 3 days. Then the Murray River firstly refilled the eastern corner of Lake Alexandrina (near to Poltalloch) within 6 days. The flow distribution at 9 days was basically controlled by the bathymetry of the lake and reached the lowest part of Lake Alexandrina near the outflow. In the following days (12–18 days), the flushing process mainly concentrated in the western part of the lake. From 21 days, the flushing process gradually turned to the northeastern part of the lake, working northwards near to the shoreline of Lake Alexandrina. This resulted from the common interaction of bathymetry and wind. In the following flushing process, the flushing trend was mainly from northeast to southwest. For Lake Albert, there was nearly no flushing process as not much lake water flowed from Lake Alexandrina to Lake Albert, and there was no other output for Lake Albert. From these results, it is clear that water quality processes will be significantly affected within the lake or reservoir and appropriate modelling studies need to be undertaken when a CR is designed with a given bathymetry and meteorological conditions.

A good inventory of over 30 existing lake water quality models is found in literature (Saloranta et al., 2004), and suitable models should be chosen and used to predict water quality variations that occur during various seasons depending on inflows, demand, flushing ability of the inflowing water, etc. If the CR is used for water supply purposes, Table 3.3 provides typical guideline values developed in several countries that can be used as an indicator of the quality that is to be expected.

3.5 Real-time monitoring of water quality

It is clear from Section 3.2 that a significant number of potential contaminants in surface waters can affect the intake water quality of a CR. Since our interest is in capturing good quality river water for an intended purpose, in situ water quality monitoring must be undertaken at selected locations. To measure water quality in real time, conventional probes can be used for temperature, whereas turbidity and conductivity can be readily used to measure sediment and salt concentration, respectively, in the water. The ion-selective electrode technique can be used to measure pH,

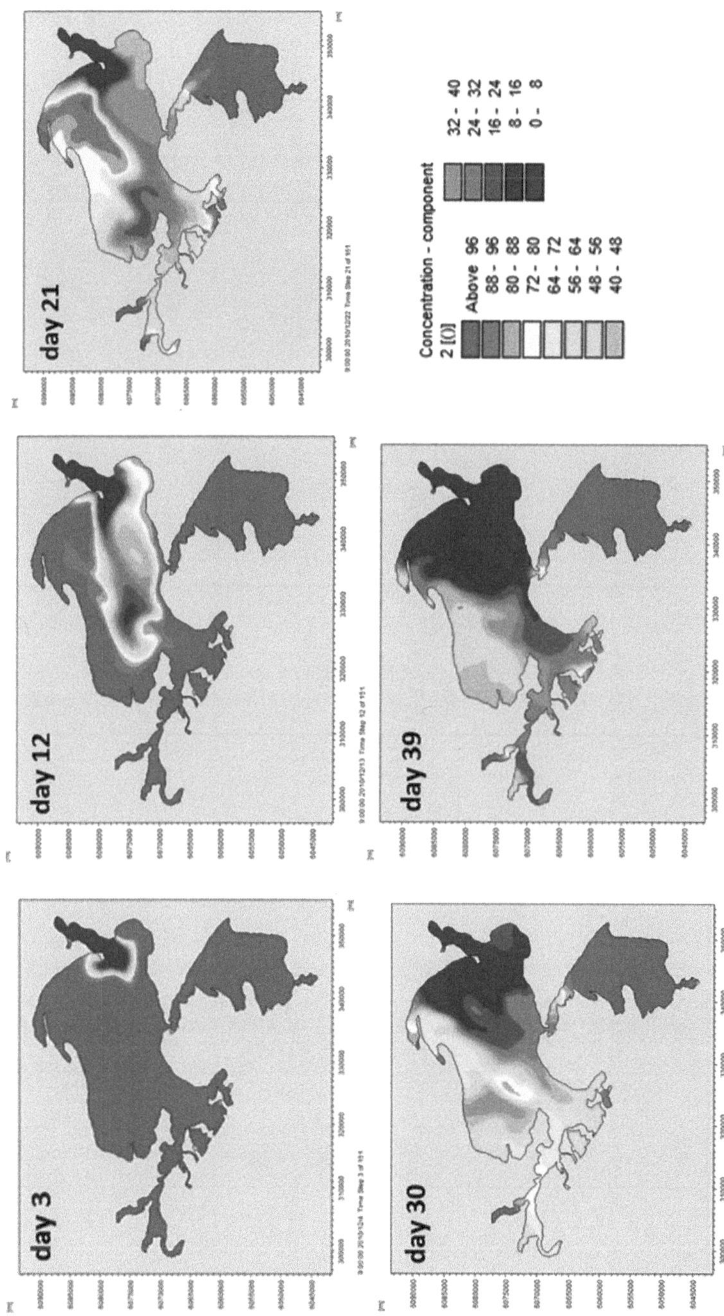

Figure 3.5 Flushing time studies of Lake Alexandrina and Lake Albert, Australia (Liu, 2017).

Table 3.3 Typical water quality guidelines of selected parameters for reservoirs used for water supply purposes.

Parameter	Unit	Australia[1]/New Zealand (ANZECC, 2000; ANZG, 2018)	Japan[2] (MOE, 2019)	UK[3] (DEFRA, 2014)	USA[4] (USEPA, 2017)	Remarks
Chemical oxygen demand	mg/L	None specified	≤ 1–3	7–11	None specified	[1]Southeast Australia, trigger values
Dissolved oxygen	mg/L	90%–110% sat	≥ 7.5	>6-9		[2] Lakes and reservoirs over 10 GL capacity; water supply classes 1 and 2
Total nitrogen	μg/L	350	670–17,00[5]	None		[3]High- to good-quality river
NH_4–N	μg/L	10		200–300		
NO_x-N	μg/L	10			10,000	
Total phosphorous	μg/L	10	5–100[5]	≤ 5–49[6]	State/location specific	[4]Protection of human health
Soluble-reactive phosphorous	μg/L	5				
Chlorophyll-a	μg/L	5	None specified	None specified	State/location specific	[5]Site-specific target values
pH		6.5–8.0	6.5–8.5	None specified	5–9	[6]Depending on alkalinity and total depth (shallow vs deep)

Total dissolved solids (salinity)	mg/L		None specified		250	[7]Additional 21 substances are being monitored
Conductivity	µS/cm	20–30		1000		
Total suspended solids	mg/L	None specified	≤ 1–5	≤ 25	None specified	[8]Additional 6 substances are being monitored
Turbidity	NTU	1–20			1–10	
Total coliform	MPN/100 mL	None specified	≤50–1000	None specified		[10]No standards or guidelines are available to date
Toxic organics	µg/L	Default guideline values are available for 135 organic and inorganic toxicants (www.waterquality.gov.au)	Standards available for 15 substances[7]	Proposed standards available for 23 priority substances	Recommended criteria for 146 substances	
Toxic inorganics	µg/L		Standards available for 9 substances[8]	Proposed standards available for 6 priority substances		
Emerging contaminants (e.g. microplastics)	—	N/A[10]				

ammonium and nitrate. Optical sensing method can be used to measure dissolved oxygen concentration. Fluorescence sensors are now becoming available to measure nutrients in real time. With the advancement in cheaper sensors and fast communication technologies such as 5G, it should now be possible to measure in real time the following water quality parameters: pH, temperature, turbidity, conductivity, DO, DOM, ammonium, nitrate, phosphorous and chlorophyll-a using a range of probes and in situ analytical methods (Blaen et al., 2016). The availability of such real-time data will be a significant bonus to the design and operation of a CR.

3.6 Conclusions

For the design of CRs, water quality considerations are vital. When river flow is considered, it can be described by volumetric flow rate, water depth and the spatial and temporal flow variability. However, when water quality is considered, the number of water contaminants such salts, sediments, nutrients, microorganisms, organic and inorganic chemical toxicants including heavy metals, biodegradable organic matter and emerging micropollutants are numerous, and their spatial and temporal variations are complex. When designing a CR, it is important to monitor selected water quality parameters in situ for several years to understand the water quality dynamics near the proposed water intake of a CR. Concurrently suitable water modelling studies need to be undertaken on the catchment river ecosystem to predict pollutant export to a CR and also model water quality processes within the CR to understand flushing as well as water age to prevent water quality deterioration. It should be noted that there are potentially a number of pre- and posttreatment processes that can be pursued to ensure that water quality is guaranteed for intended use. However, consideration of various water treatment processes is beyond the scope of this chapter.

References

Akdogan, Z., Guven, B., 2019. Microplastics in the environment: a critical review of current understanding and identification of future research needs. Environmental Pollution 254, 113011.

Ambrosetti, W., Barbanti, L., Sala, N., 2003. Residence time and physical processes in lakes. Journal of Limnology 62 (Suppl. 1), 1–15.

ANZECC, 2000. Australian and New Zealand Guidelines for Fresh and Marine Water Quality, vol. 1. The guidelines/Australian and New Zealand Environment and Conservation Council, Agriculture and Resource Management Council of Australia and New Zealand, Canberra, Australia.

ANZG, 2018. Australian and New Zealand Guidelines for Fresh and Marine Water Quality (Accessed 22 December 2019). https://www.waterquality.gov.au/guidelines/anz-fresh-marine.

ASTM, 2007. Standard Test Method for Determining Sediment Concentration in Water Samples. D3977-D39797R07. West Conshohocken, PA.

Blaen, P.J., Khamis, K., Lloyd, C.E.M., Bradley, C., Hannah, D., Krause, S., 2016. Real-time monitoring of nutrients and dissolved organic matter in rivers: capturing event dynamics, technological opportunities and future directions. Science of the Total Environment 569, 647–660.

Casatta, N., Stefani, F., Pozzoni, F., Guzzella, L., Marziali, L., Mascolo, G., Viganò, L., 2016. Endocrine-disrupting chemicals in coastal lagoons of the Po River delta: sediment contamination, bioaccumulation and effects on Manila clams. Environmental Science and Pollution Research 23 (11), 10477–10493.

Chin, D.A., Sakura-Lemessy, D., Bosch, D., Gay, P.A., 2009. Watershed-scale fate and transport of bacteria. Transactions of the ASABE 52.

DEFRA, 2014. Water Framework Directive implementation in England and Wales: new and updated standards to protect the water environment. Department for Environment, Food and Rural Affairs 41. Llywodraeth Cymru Weslch Governement.

Ellison, C.A., Savage, B.E., Johnson, G.D., 2014. Suspended-sediment concentrations, loads, total suspended solids, turbidity, and particle-size fractions for selected rivers in Minnesota, 2007 through 2011. U.S. Geological Survey Scientific Investigations Report 2013–5205 43.

Fu, B., Merritt, W.S., Croke, B.F.W., Weber, T.R., Jakeman, A.J., 2019. A review of catchment-scale water quality and erosion models and a synthesis of future prospects. Environmental Modelling and Software 114, 75–97.

Gong, W., Shen, J., Hong, B., 2009. The influence of wind on the water age in the Tidal Rappahannock River. Marine Environmental Research 68 (4), 203.

Groten, J.T., Johnson, G.D., 2018. Comparability of river suspended sediment sampling and laboratory analysis methods. U.S. Geological Survey Scientific Investigations Report 2018–5023, 23.

Im, S., Brannan, K.M., Mostaghimi, S., Kim, S.M., 2007. Comparison of HSPF and SWAT models performance for runoff and sediment yield prediction. Journal of Environmental Science and Health, Part A 42 (11), 1561–1570.

Koelmans, A.A., Mohamed Nor, N.H., Hermsen, E., Kooi, M., Mintenig, S.M., De France, J., 2019. Microplastics in freshwaters and drinking water: critical review and assessment of data quality. Water Research 155, 410–422.

Lee, J., Lee, S., Yu, S., Rhew, D., 2016. Relationships between water quality parameters in rivers and lakes: BOD_5, COD, NBOPs, and TOC. Environmental Monitoring and Assessment 188 (4), 252.

Liu, J., 2017. Hydrodynamic and Salinity Simulation in the Lower Lakes, South Australia and Proposed Coastal Reservoir. PhD Thesis. University of Wollongong, Australia.

Liu, J., Sivakumar, M., Yang, S., Jones, B.G., 2018. Salinity modelling and management of the lower lakes of the Murray–Darling basin, Australia. WIT Transactions on Ecology and the Environment 228, 257–258.

López-Doval, J.C., Montagner, C.C., de Alburquerque, A.F., Moschini-Carlos, V., Umbuzeiro, G., Pompêo, M., 2017. Nutrients, emerging pollutants and pesticides in a tropical urban reservoir: spatial distributions and risk assessment. The Science of the Total Environment 575, 1307–1324.
May, D., Sivakumar, M., 2008. Comparison of artificial neural network and regression models in the prediction of urban stormwater quality. Water Environment Research 80 (1), 4–9.
May, D., Sivakumar, M., 2009a. Prediction of nutrient concentrations in urban storm water. Journal of Environmental Engineering 135 (8), 586–594.
May, D.B., Sivakumar, M., 2009b. Prediction of heavy metal concentrations in urban stormwater. Water and Environment Journal 23 (4), 247–254.
McNeil, V.H., Cox, M.E., 2000. Relationship between conductivity and analysed composition in a large set of natural surface-water samples, Queensland, Australia. Environmental Geology 39 (12), 1325–1333.
MDBA, 2013. Approach for Estimating Salt Export from the River Murray System to the Southern Ocean. Murray-Darling Basin Authority. Australia Publication No 15/13.
MOE, 2019. Environmental Quality Standards for Water in Japan. Ministry of Environment Japan (Accessed December 2019). https://www.env.go.jp/en/water/wq/wp.pdf.
Pantelaki, I., Voutsa, D., 2019. Organophosphate flame retardants (OPFRs): a review on analytical methods and occurrence in wastewater and aquatic environment. Science of the Total Environment 649, 247–263.
Peña-Guzmán, C., Ulloa-Sánchez, S., Mora, K., Helena-Bustos, R., Lopez-Barrera, E., Alvarez, J., Rodriguez-Pinzón, M., 2019. Emerging pollutants in the urban water cycle in Latin America: a review of the current literature. Journal of Environmental Management 237, 408–423.
Rice, E.W., Baird, R.B., Eaton, A.D., 2017. Standard Methods for the Examination of Water and Wastewater, twenty third ed. American Public Health Association, American Water Works Association, Water Environment Federation.
Saloranta, T.M., Malve, O., Bakken, T.H., Ibrekk, A.S., 2004. Lake Water Quality Models and Benchmark Criteria. Delivery Report from the Lake Model Work Package (WP6) of the BMW-Project.
Smith, H.G., Sheridan, G.J., Lane, P.N.J., Nyman, P., Haydon, S., 2011. Wildfire effects on water quality in forest catchments; a review with implications for water supply. Journal of Hydrology 396 (1–2), 170–192.
Snow, D.D., Cassada, D.A., Biswas, S., Malakar, A., D'Alessio, M., Carter, L.J., Johnson, R.D., Sallach, J.B., 2019. Detection, occurrence, and fate of emerging contaminants in agricultural environments (2019). Water Environment Research 91 (10), 1103–1113.
Tijani, J., Fatoba, O., Babajide, O., Petrik, L., 2016. Pharmaceuticals, endocrine disruptors, personal care products, nanomaterials and perfluorinated pollutants: a review. Environmental Chemistry Letters 14 (1), 27.
Troester, M., Brauch, H.-J., Hofmann, T., 2016. Vulnerability of drinking water supplies to engineered nanoparticles. Water Research 96, 255–279.
USEPA, 2017. Water Quality Standards Handbook: Chapter 3: Water Quality Criteria. EPA-823-B-17-001. US Environmental Protecion Agency, Office of Water, Office of Science and Technology, Washington, DC (Accessed November 2019). https://www.epa.gov/sites/production/files/2014-10/documents/handbook-chapter3.pdf.
Warziniack, T., Sham, C.H., Morgan, R., Feferholtz, Y., 2017. Effect of forest cover on water treatment costs. Water Economics and Policy 03 (04), 1750006.

Whitehead, P.G., Jin, L., Bussi, G., Voepel, H.E., Darby, S.E., Vasilopoulos, G., Manley, R., Rodda, H., Hutton, C., Hackney, C., Van Pham Dang, T., Hung, N.N., 2019. Water Quality Modelling of the Mekong River Basin: Climate Change and Socioeconomics Drive Flow and Nutrient Flux Changes to the Mekong Delta, pp. 218–229.

Withers, P.J.A., Jarvie, H.P., 2008. Delivery and cycling of phosphorus in rivers: a review. Science of the Total Environment 400 (1), 379–395.

Wuest, A., Lorke, A., 2003. Small-scale hydrodynamics of lakes. Annual Review of Fluid Mechanics 35 (1), 373.

Xu, Z., Ma, J., Wang, H., Hu, Y., Yang, G., Deng, W., 2018. River discharge and saltwater intrusion level study of Yangtze River estuary, China. Water 10 (6), 683.

Yadav, M.K., Short, M.D., Aryal, R., Gerber, C., van den Akker, B., Saint, C.P., 2017. Occurrence of illicit drugs in water and wastewater and their removal during wastewater treatment. Water Research 124, 713–727.

CHAPTER 4

Geotechnical considerations for coastal reservoirs

T.G. Sitharam[1,2], Sreevalsa Kolathayar[3]
[1]Department of Civil Engineering, Indian Institute of Science, Bangalore, Karnataka, India; [2]IIT Guwahati, Guwahati, Assam, India; [3]Department of Civil Engineering, National Institute of Technology Karnataka, Surathkal, Karnataka, India

4.1 Introduction

About 71% of the earth surface is covered by water, in which ocean covers about 97% and the remaining 3% is freshwater. Of this 3%, 69% is held in the form of glaciers and ice caps. The total amount of freshwater effectively accessible is merely 0.014%. The increased population, overexploitation of groundwater, pollution, deforestation, etc., have resulted in a further reduction in the availability of freshwater. The World Economic Forum has listed water scarcity as the most significant global risk that is going to impact the world in the next decade. Studies showed that, if the current trend is continuing, humanity will face a water crisis with demand exceeding the supply by 40% in 2030. The increase in population coupled with a change in lifestyle and land use has resulted in intensive exploitation of groundwater sources. Deforestation, increased number of concrete buildings and other development activities have increased the area of land covered with concrete and other paving materials. However, this has increased surface runoffs and decreased the amount of water infiltrating into the ground. In India, from the total rainfall, only 17% is utilized, remaining water flows into the ocean. Hence, there is a need for structures to store this rainwater. This can be done by 1. construction of surface water storage tanks, dams, reservoirs, etc., which will augment groundwater regionally, 2. rainwater harvesting structures like check bunds, infiltration wells, injection wells, etc., 3. underground dams (Subsurface dams) and 4. coastal reservoirs.

Geotechnology is the application of engineering concepts and new innovative geosynthetic materials along with soil or rock for the solution of engineering problems. Water through precipitation is first received by the earth, which either percolates into the ground or runoff into surface water bodies. In addition to storage of water in dams and surface water bodies, a significant portion of water is stored underground, which is

Sustainable Water Resource Development Using Coastal Reservoirs
ISBN 978-0-12-818002-0
https://doi.org/10.1016/B978-0-12-818002-0.00004-6

popularly known as groundwater. A considerable amount of rainwater is lost as runoff without being percolated into the ground because of rapid urbanization, siltation of river courses and tanks and deforestation. Groundwater depletion is a major challenge faced throughout the world in recent times. Depletion of groundwater is the primary cause for drying of natural freshwater bodies which in turn affects the water supply in a region. This needs to be addressed at present to ensure water security in the future. Geotechnics and hydrology are closely linked topics, wherein application of geotechnology based applications to control the flow of water; augment storage in tanks, ponds and dams and increase the groundwater penetration through scientific methods using the knowledge of earth materials with innovations in geotechnical engineering. Geotechnology has a major role to play in recharging the groundwater by arresting the surface runoff, thereby allowing the water to percolate into the ground. The artificial techniques for groundwater recharge can be classified as surface methods and subsurface methods. Groundwater hydrology, soil/rock characteristics and permeability are the major factors governing the efficiency of groundwater recharge techniques. Surface methods enhance groundwater infiltration by providing more residence time with the help of structural and nonstructural measures such as percolation tank, flooding, stream augmentation, ditch and furrow system and contour bunds. Afforestation is another efficient surface method that facilitates a significant amount of water infiltration. Subsurface structures like recharge wells, dug wells, infiltration wells, pits and shafts, etc., can recharge water directly to the aquifer. The assimilation capacity of the soil and hydrogeological factors must be monitored for implementation of surface recharge methods considering the depth of the aquifer, storage capacity, porosity of soil underground, permeability of the soil, land use and topography, economic aspects and degree of public acceptance (Goldberg and Gazda, 1984).

The coastal freshwater reservoir is a new emerging concept of storing floodwater in the sea close to the shoreline. The coastal reservoir can be constructed in shallow waters at appropriate locations close to the mouth of the river along with a barrage at one or two ends. Sea walls or breakwaters with some modifications along with new and sustainable construction technologies are good enough to construct the sea-based reservoirs/coastal reservoirs. This chapter presents the geotechnical consideration for the coastal reservoir to impound the floodwaters.

4.2 Innovative materials for construction and lining

This section reviews various materials of geotechnical relevance that can be used for the construction of the barriers to check water flow and enhance infiltration. With ongoing research, several innovative materials have been developed all over the world, which can serve the purpose of separation, filtration, reinforcement and drainage. Some of them are listed below:

- Tarpaulin sheets – HDPE with tar felt
- Waterproofing tar felt
- LDPE
- HDPE
- Geosynthetic clay liner)
- Grouting – cement or bentonite cutoff walls
- Cement–bentonite slurry
- Pulverized fuel ash
- Sodium-enriched bentonite

Sharma and Kozicki (1988) briefly reviewed various liner materials, their properties and their applications to prevent the spread of contaminant into groundwater. In the first case, reinforced chlorinated polyethylene liner was used to protect groundwater at a petrochemical plant. Reinforcement is provided to resist uplift pressure at the base of the liner due to the hydrostatic pressure. In the second case, the soil–bentonite liner was used to control seepage in aerated lagoon facilities constructed at Wawa, Ontario, Canada. A mixture of in situ soil and bentonite was found to be more economical than polyethylene liner in this case. Two laboratory permeability tests were conducted with soil and different percentage of sodium bentonite. The design of soil–bentonite liner was based on the permeability test done on various soil mixtures. The mixture with least laboratory permeability value was used, and the design thickness of the liner was fixed by comparing the seepage rates through the liner with the allowable seepage rate. They recommend that, in addition to adequate testing, special care should be taken during the construction of these liners (Sharma and Kozicki, 1988). Iravanian and Bilsel (2016) studied the volume change behaviour of sand–bentonite liner and the effect of cement enhancement. Mixtures of 15% bentonite–85% sand, 10% bentonite–90% sand, 5% bentonite–5% cement–90% sand and 15% bentonite–5% cement–80% sand were prepared. From the compaction test results, they have concluded that the addition of more bentonite increases optimum moisture content and maximum dry density. The volume change behaviour of samples containing cement was recorded as very low in both swell and consolidation parts.

4.3 Groundwater dams or subsurface dikes

A groundwater dam (Fig. 4.1) is a wall below the ground across a watercourse, which prevents the natural flow of groundwater and stores water below the ground surface to meet future demands. The arrest of the flow of groundwater out of the subbasin increases the storage within the aquifer. It enables the rising of water levels in upstream of groundwater dam and saturates the otherwise dry part of the aquifer. These dikes can capture groundwater flowing even near the surface of the ground, in dry watercourse beds, and the water can be extracted via wells on top of the dam. The subsurface dam does not cause submergence of land as the water is stored below the ground within an aquifer. The area above the barrier can be utilized, which is not possible in surface dams. Also, the underground dams do not face evaporation loss from the reservoir, and siltation in the reservoir is prevented. Further, the potential disaster like the collapse of dams can be avoided with the introduction of underground dams.

A subsurface dike is defined as a system to store groundwater by a 'cutoff wall' (dam body), set up across a groundwater channel. These subsurface dams are also known as subsurface dikes and underground dams. The subsurface dams are constructed at some depth (approximately 1 m) below

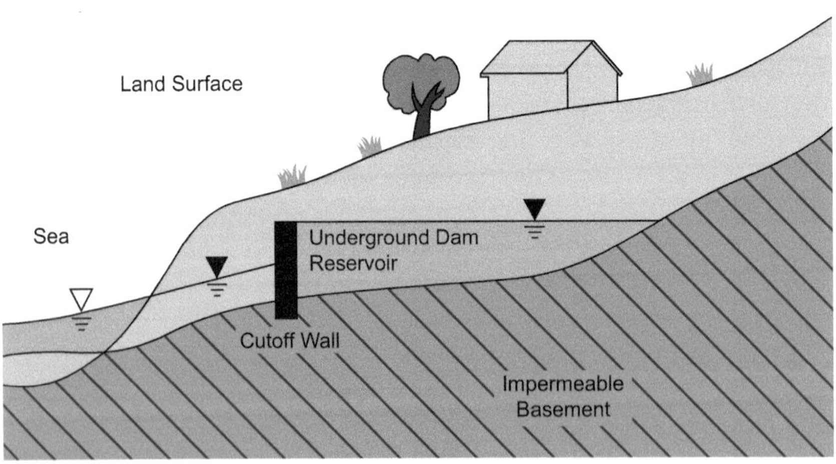

Figure 4.1 Schematic of a groundwater dam. *(Image source: http://www.aquor.com.au.)*

the riverbed, as shown in Fig. 4.1. As rain occurs, this cutoff wall will prevent the downstream flow of infiltrated water to the ocean and make it available in the upstream of the dam. This water can be accessed via wells from the upstream of the dam and surrounded preferably by gravel infiltration galleries.

A large number of subsurface dams have been already constructed in different parts of the world. Brazil, Kenya and Japan are the pioneers in subsurface dam construction, and they have built a large number of subsurface dams. During the period 1960–2010, both Kenya and Brazil have constructed around 500 dams, Japan has made about 21 dams and in India, only 5 dams have been built. Japan has created quite a good number of subsurface dams with length varying from 50 to 2500 m and height ranging from 4 to 80 m.

4.3.1 Classification of subsurface dams

Subsurface dams can be classified into two categories depending upon their purpose and material used for construction. Based on purpose, it can be classified as the dam-up type and saltwater intrusion prevention type (Fig. 4.2). Dam-up type dam, which can be further categorized as storage type and runoff control type dam, is constructed inland and has the

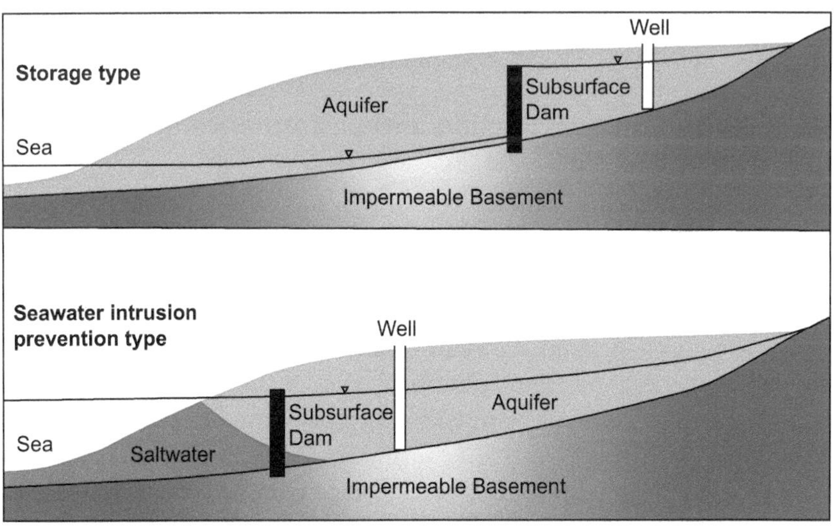

Figure 4.2 Classification of subsurface dams. *(Image source: Raju, N.J., Reddy, T.V.K., Muniratnam, P., Gossel, W., Wycisk, P., 2013. Managed aquifer recharge (MAR) by the construction of subsurface dams in the semi-arid regions: a case study of the Kalangi river basin, Andhra Pradesh. Journal of the Geological Society of India, 82 (6), 657–665.)*

potential to store plenty of groundwater. These types of dams store groundwater and prevent its flow, thereby increasing the groundwater level. Saltwater intrusion prevention type dam prevents intrusion of saltwater to the freshwater, thereby safeguarding the available freshwater resources. These dams are usually constructed near the coast to check seawater intrusion into groundwater aquifers.

For the construction of subsurface dams, a variety of materials can be used. The materials used for construction include concrete, brick masonry, stone masonry and grout injection screens.

4.3.2 Benefits and demerits

Subsurface dams have numerous advantages compared to conventional surface dams. The main advantage of a subsurface dam is that it recharges the groundwater level. The water stored in the upstream side of a subsurface dam is less susceptible to pollution. Therefore, the quality of the water will be good. As the water is stored, some depth below the ground level evaporation loss will be less (almost zero). Construction of surface dams causes submergence of the upstream area and results in social and environmental issues. There is no such problem of upstream submergence in the case of a subsurface dam. The water stored in the upstream side of the subsurface dam can be extracted by installing well, which requires pumping. This is an increased operation cost, which is a disadvantage compared with surface dams. The advantages and disadvantages of subsurface dams are listed in Table 4.1.

4.3.3 Studies on construction and performance of subsurface dams

This section highlights studies conducted on various subsurface dams around the world, their construction, performance, environmental impact and capability of solving water scarcity problems. Khairi et al. (2010) have done a 3D finite element numerical analysis of the subsurface concrete dam for the Holy city of Makkah, Saudi Arabia. The city has 1.4 million permanent residents, and the population exceeds 3 million during the pilgrimage season. 90% of the water supply comes from desalination plants, which may get interrupted due to unanticipated circumstances. Hence, a subsurface dam was proposed as a water storage method for the Holy city of Makkah. A place called WadiNaman was considered as the best location. The proposed dam has a total length of 912.87 m. A plastic concrete cutoff wall was recommended to create the impermeable barrier. This study

Table 4.1 Advantages and disadvantages of the subsurface dam.

Advantages	Disadvantages
• Recharges groundwater • Evaporation loss is relatively less • Water storage without submergence of land • Quality of water obtained will be good depending on the soil geochemistry • Susceptibility to pollution is less • Will not obstruct the natural river course • Requires low initial cost • Maintenance cost will be less • Locally available materials can be used • Destruction of the dam will not cause any harm • River ecology remains unchanged	• Operating cost required for pumping is more than that of subsurface dams • Accurate estimation of the storage water volume and groundwater recharge is difficult • Requires low permeability rock at the base and sides • Potentially higher evapotranspiration loss

describes the behaviour of cutoff walls under loading. They have concluded that maximum horizontal and vertical stresses developed at about 1/6th height of the wall.

Raju et al. (2006) have presented a case study of subsurface dams constructed to harvest rainwater in the Swarnamukhi river basin, Andhra Pradesh, India. The dam was proposed to rejuvenate the river basin, which is essentially a semiarid tract with an average annual rainfall of 1000 mm. By studying topographic and geological maps and based on depth to rock, dam alignment was fixed. Four subsurface dams were constructed with puddled clay 0.9 m thick which has been keyed into a hard-end connection on either side of the riverbank. The top surface of the dam has been limited to 1 m below the riverbed level. The dam has been keyed into weathered bedrock 0.55 m below bedrock surface. The clay wall was compacted with wooden rammers to obtain proctor density. A 200 μm thick low-density polyethylene film was emplaced on the upstream, downstream and top of the dam to retain moisture content and protect the clay from yielding to cracks. The well hydrograph of these subsurface dams shows that the water level fluctuation of 3.1–10 m before construction of the barrier has been reduced to 0.4–3.1 m after the construction of barrier.

Ramasesha et al. (2002) have studied the efficiency of subsurface dikes as a groundwater conservation structure. Based on the study, subsurface dikes constructed on an experimental basis by Central Ground Water Board in hard rock terrain of Tamil Nadu, India, have proved to be effective. In this study, the details of subsurface dike constructed at Nallam Pillai Petral are given. For the construction of the dike, a trench was excavated up to bedrock, and a brick wall was raised to 1.5 m below ground level. The trench was then backfilled with clay. Observation wells were constructed on both upstream and downstream to monitor the impact of the dike on the water level. From the observation wells on the upstream side, it was found that there was a rise in water level after the construction of dike. The observation wells in the downstream side did not show any decline in water level after the construction of the dike. Based on the study, they have concluded that the subsurface dikes are most suitable at locations where hard rock is encountered within a depth of 8–10 m below ground level.

Ishida et al. (2003) have reviewed the basics of subsurface dams, its construction and the problems associated with the sustainable use of groundwater. In this review paper, they have described the construction of subsurface dams around the world during different periods. Based on the statistics given in the paper, until the 1990s, the number of subsurface dams constructed in different countries was less, and their storage capacities were small. But after 1990, the mixed-in-place slurry wall method made it possible to construct subsurface dams that can store 1 million m^3 or more of groundwater. The study of the impact of subsurface dams on groundwater has shown that the presence of saline soils will cause significant levels of salination. Hence, they recommend choosing sites without saline soils for the construction of subsurface dams.

Fazal et al. (2005) have studied the construction of subsurface dams and their environmental impacts. The study focused on the two subsurface dams constructed in Miyakojima Island, Japan – Sunagawa and Fukuzato. The construction of these dams started in 1987 and was completed by 2001. The total capacity of these two dams is about 20,000,000 m^3, making this project one of the biggest subsurface dam systems in the world. The cutoff walls were built with the in situ churning method. The groundwater is drawn by 147 tube wells. The first concern of this study is the presence of 'infiltrated clay', which affects the permeability of the in situ soil. The island consists of Quaternary Ryukyu limestone with high permeability. Ryukyu limestone is covered with lateritic clay of 0.5–1 m thickness. They have found that infiltrated rain produces several caves within the Ryukyu limestone. Once these caves start to conduct soil along with water, the intruded soil called 'infiltrated clay' closes the pores of the limestone, thus decreasing its hydraulic conductivity. The intake wells were arranged based on the relation between infiltrated clay content and hydraulic conductivity. The second concern was the impact of the subsurface dam on groundwater nitrate concentration. From observations, they have found that the impact of cutoff walls on the underground environment is too small to influence the nitrate concentration.

Onder and Yilmaz (2005a,b) defined underground dams as a tool for sustainable development and management of groundwater resources. In their study, a review on groundwater dams, both sand dams and subsurface dams has been presented. Also, the design and construction of subsurface dams and various materials that can be used as the barrier are also discussed. MODFLOW has been used for the numerical simulation of groundwater flow. For this study, numerical modelling of a hypothetical case and that of Çamlı Köyü, Turkey, has been done. The groundwater flows under the natural condition and after inserting the barrier have been simulated. The variation of water table elevation is obtained as a contour line map from MODLOW simulation. Based on the slope of the hydraulic head across the slope, the pumping rate has been fixed to avoid saltwater intrusion. It has been concluded that the development of the groundwater basin and the concept of storing water in subsurface dams is an attractive solution for the water shortage problem because of its relatively low social and environmental impacts.

Suseela (2005) has reported a study of low-cost subsurface dike using bentonite clay across Bharatapuzha, Kerala. As described in this paper, the unit land of Kerala receives 2.5 times more rainfall compared to the national average. Due to steep topography, most of this rainwater is not retained, and only 5% of runoff is believed to be converted into groundwater. Subsurface dams were proposed across Bharatapuzha at 14 locations, to reduce the water scarcity problems in the districts of Thrissur, Malappuram and Palakkad during summers. For a subsurface dike constructed at Thrangali Kadavu in Palakkad district across Bharatapuzha, the site was considered suitable because it can avail water for irrigation (a lift irrigation scheme is situated approximately 105 m upstream of the proposed weir line) and drinking purpose (a drinking water scheme is also situated 1 km upstream of subsurface dike). Sand–bentonite–cement grout was used to construct the impermeable layer. The performance of the subsurface dike was studied by monitoring the water level in the adjacent wells at left and right banks of the river, and it was found that the water level increased during summer by about 0.6 m.

4.3.4 Benefits of storing water in the sand

There are several benefits of storing water in the sand in an underground dam as sand is a type of soil with unique characteristics in terms of permeability, drainage and void space:

- Empty spaces between sand (voids) are filled when a dry riverbed is flooded.
- Tiny voids get saturated slower, and less water can be extracted compared to coarse material.
- Low evaporation losses.
- No contamination from livestock or other animals.
- No waterborne diseases from mosquitoes or reptiles.

It is observed that coarser the sand, the higher the volume of water that can be extracted. Table 4.2 presents the water extraction potential of underground dam based on the particle size of the soil below the surface.

4.3.5 Potential riverbeds to construct a subsurface dike

It is important to assess the feasibility of riverbed before constructing a subsurface dike. Every riverbed cannot reap the benefits of underground dams as many factors affect the storage of water below the ground. The geography, topography, hydrology and geology are deciding factors for water storage underground. Usually, subsurface valleys are formed where the ground is undulating. The riverbank must be well defined and stable to construct a subsurface dam. Soil with large porosity, which can hold large volumes of water, including the alluvial gravel layer, limestone and volcanic rocks, are very favourable. A minimum topographic riverbed slope is required with a preferred range of gradients between 0.2% and 4%. Other factors to be considered before construction of a subsurface dam are technical aspects and social aspects as listed below.

Table 4.2 Water extraction potential in terms of particle size.

	Silt	Fine sand	Medium sand	Coarse sand
Size (mm)	<0.5	0.5–1.0	1.0–1.5	1.5–5.0
Saturation	38%	40%	41%	45%
Water extraction	5%	19%	25%	35%

4.3.5.1 Technical aspects

- Location and types of trees indicating water
- Location of waterholes
- Topographic map
- Geological maps
- Rainfall, river discharge, groundwater level, water quality
- Location and types of rocks and boulders
- Location of calcrete (this turns saline in water)
- Coarseness of sand
- Location of hand-dug wells, boreholes and weirs

4.3.5.2 Social aspects

- Names of houses, schools and road crossings
- Ownership/number of beneficiaries
- The type and population of livestock served
- Land ownership issues

Technical consultation of skilled practising engineers and public consultation are necessary to incorporate technical and social aspects efficiently.

4.4 Geotechnics for coastal reservoirs

The major component of geotechnology in coastal reservoirs is the construction and design of sea dike. A sea dike is an embankment used to protect low-lying areas against seawater inundation. Storage of the abundant monsoon water can be done close to the coast using sea dikes, which otherwise runs off to the ocean. Coastal reservoirs are bounded by impermeable sea dikes at one side and the coast on the other side (Yang, 2015). Sea dike is a static feature, and it will conflict with the dynamic nature of the coast and impede the exchange of sediment and saltwater between land and sea at the mouth of the river. Sea dikes may also be constructed from a variety of materials, most commonly geosynthetic tubes, geocells, reinforced concrete, boulders, steel or gabions. Sea dikes are primarily used at exposed coasts, but they are also used at moderately exposed coasts. In the case of coastal reservoirs, sea dikes are used for the separation of ocean saltwater from the floodwater from rivers stored in coastal reservoirs (Sitharam, 2018; Kolathayar et al., 2019).

4.4.1 Typical sea dike cross-sectional design

Sea dike cross-sectional design includes crest level, cross-sectional dimensions, crest structure, dike body and dike toe, which fulfil the technical and economic requirements. Design cross sections of the dike are selected on the basis of geological conditions, materials used for the construction of dike, filling materials, external forces, the layout of the dike and also the operational requirements. Sloping dikes, wall-type dikes and composite dikes are three different types of sea dikes based on geometrical shape. The design of the dike crest level, dike body, filter layers, slope protection layers (both on seaside and landside/coastal reservoir side) and toe protection is important. In addition to storage of water, dikes reduce the amount of energy dissipated by the waves reaching the coastlines. It protects the coastline from the tidal action and provides coastal defence. Further, it prevents the erosion of the coastline and creates calm water in the coastal reservoir area for various activities. The most important function of the dike is to separate the saltwater and the freshwater in the coastal reservoir. It facilitates freshwater fishing and other activities in calm water conditions, and it can also provide dock or quay facilities along with the support of floating solar panels for energy production. The main features of the sea dike can be seen in Fig. 4.3.

4.4.2 Impermeable sea dike for coastal reservoir

A freshwater reservoir located in the sea close to the mouth of a river is bounded by sea dike to store the required quantity of river floodwater. Flood water is a natural resource, and water quality is comparable to drinking water. Sea dike that bounds the coastal reservoir shall be impermeable so that saltwater on the seaside does not mix with the freshwater stored in the coastal reservoir. The wall with barrier lining is designed such that it will prevent mixing of surrounding seawaters with the impounded

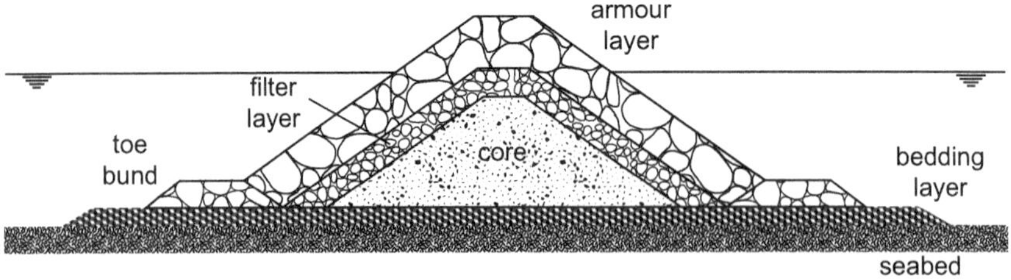

Figure 4.3 Typical features of sea dike.

floodwaters. An effective impermeable barrier between the freshwater (floodwater) and the salty seawater is needed to do the following:

- Separation of clean river water from saltwater.
- Protection of collected freshwater against external pollution.
- Prevention of saltwater intrusion into the stored freshwater.

A typical cross section of the sea dike for the coastal reservoir is suggested in Fig. 4.4.

4.4.3 Bathymetric profile

For the construction of sea dike for coastal reservoirs, it is important to generate the bathymetry profile of the sea bed. It will enable to calculate the storage volume of coastal reservoirs of different dimensions. The General Bathymetric Chart of the Oceans (GEBCO) is a publicly available bathymetric chart of the world's oceans. The project was conceived with the aim of preparing a global series of charts showing the general shape of the seafloor. Over the years, it has become a reference map of the bathymetry of the world's oceans for scientists and others. Since 1903, the GEBCO Project has been collecting bathymetric data and mapping the earth's oceans. GEBCO is an IHO and IOC joint Project that relies largely on the voluntary efforts of an international collaborating community of scientists and hydrographers with the support of their parent organizations. The bathymetry profile for Netravati basin was created by defining the datum for the file, carrying out the projections process for the region using the Decker 1986 north zone 44N, using GIS applications.

A bathymetric chart is the submerged equivalent of an above-water topographic map. Bathymetric charts are designed to present accurate, measurable description and visual presentation of the submerged terrain. In an ideal case, the joining of a bathymetric chart and topographic map of the

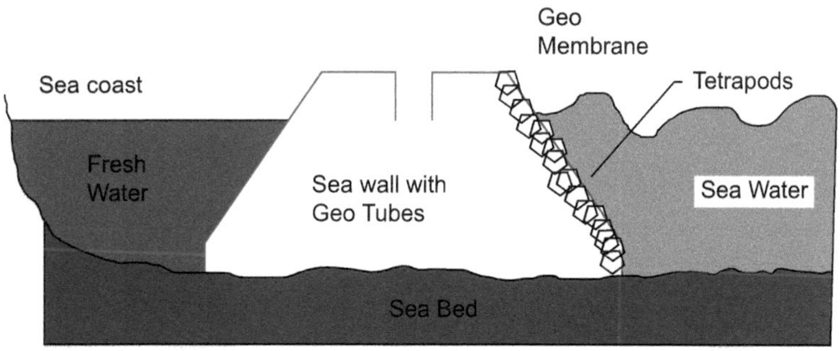

Figure 4.4 A typical impermeable sea dike for the coastal reservoir.

same scale and projection of the same geographic area would be seamless. The only difference would be that the values begin increasing after crossing the zero at the designated sea-level datum. Thus, the topographic map's mountains have the greatest values, while the bathymetric chart's greatest depths have the greatest values. Simply put, the bathymetric chart is intended to show the land, if overlying waters were removed in exactly the same manner as the topographic map of the seafloor.

Through the use of detailed depth contours and full use of bathymetric data, the size, shape and distribution of underwater features are vividly portrayed. No other map or chart gives this descriptive picture of the ocean bottom terrain with size, shape and distribution of underwater features. The bathymetric map serves as the basic tool for performing scientific, engineering, marine geophysical and environmental studies that are required in the development of energy and marine resources. A typical bathymetric profile of the Arabian seabed near the mouth of the Netravati River is shown in Fig. 4.5. The lithological sequence of a typical borehole from an offshore site in the Arabian Sea is presented in Fig. 4.6.

4.4.4 Sea dike construction

The geotextile mattress or geomat method suggests stacking several layers of flat geotextile mattresses for dike construction. It provides better lateral

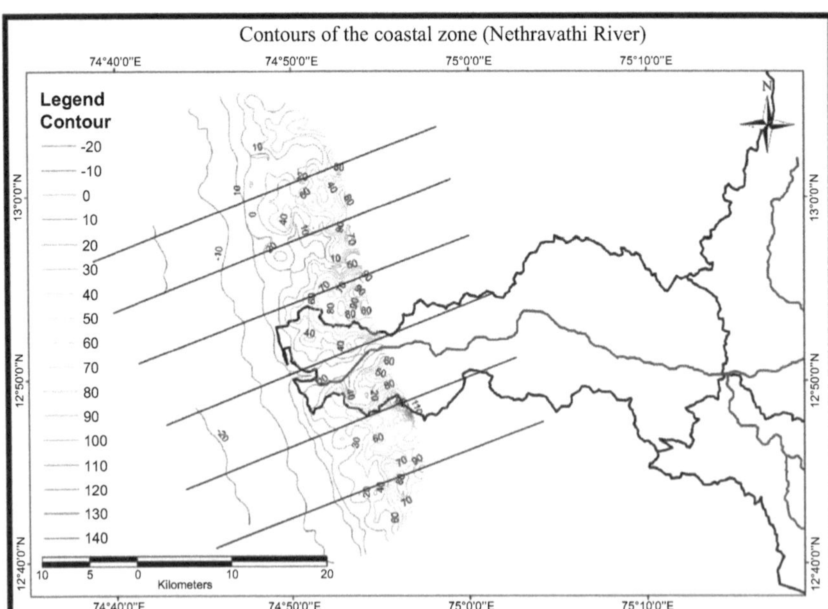

Figure 4.5 Bathymetry of the ocean bed near the Netravati River.

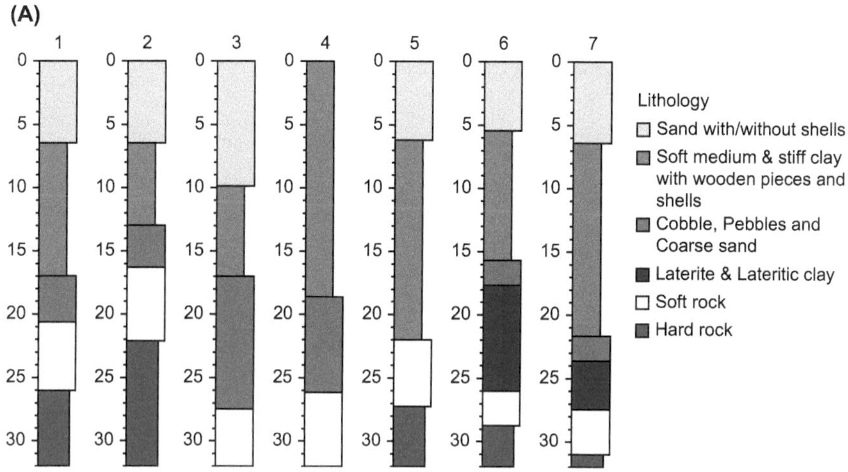

Figure 4.6 Stratigraphy/Lithological Sequence of typical Boreholes from an offshore site in the Arabian Sea. *(Image source: Parthasarathy, C.R., Sitharam, T.G., Kolathayar, S., 2019. Geotechnical considerations for the concept of coastal reservoir at Mangaluru to impound the flood waters of Netravati River. Marine Georesources and Geotechnology, 37 (2), 236–244.)*

stability as compared to the geotextile tube method and is suitable for the construction of breakwater or dikes in shallow water (Chu et al., 2012). In rough seawater, prefabricated concrete caisson will be a better option for the sea dike, but ground improvement may be necessary for the weak seabed. The suction caisson method is good for dikes constructed in deep sea, and improvement of the weak seabed is not necessary with the use of suction skirt. The dike construction using geotubes filled with sand is becoming popular, and it has added advantage of using locally available sand for the construction. The sand filled in geotubes is very good in strength as well. This method emerges to be one of the efficient and cost-effective methods of dike construction.

The dike construction using geotubes filled with sand (Fig. 4.7) is one of the modern methods used. Initially, the geotubes are filled with air or water in which the height can be changed. But in recent times, the geotubes are filled with sand or dehydrated soil. For nearshore or offshore project, a suction dredger can be used to pump sand from the sea bed or a sandpit directly into the geotextile tubes. The horizontal pressure taken up by the geotubes are very less, and therefore to overcome this disadvantage of geotubes, geomats are used. Conventionally, rock revetments were used for erosion protection on both sides, which is now replaced by geotubes.

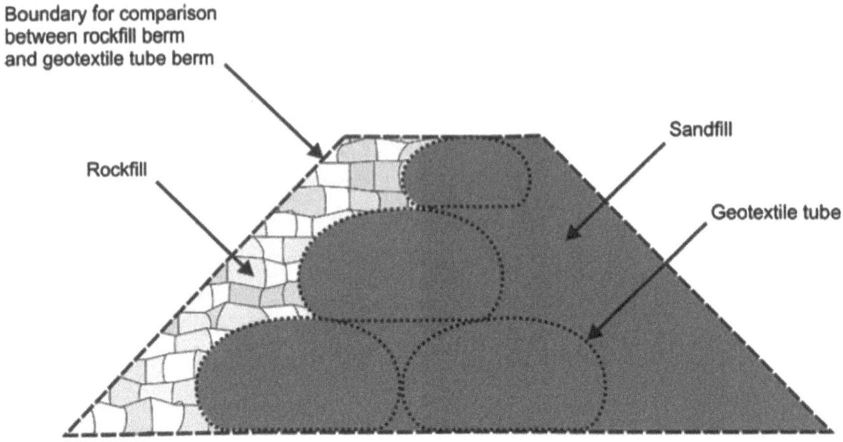

Figure 4.7 Dike construction using geotubes filled with sand. *(From Yee, T.W., Lim, L.K., Ter Harmsel, M., Choi, J.C., Hwang, S.P., 2013. Geotextile tubes as a replacement for rockfill for the polder dike in Saemangeum, South Korea, Terra et Aqua (133), 10–23.)*

The real-life example of a reservoir built with dike is the Saemangeum seawall in South Korea. The construction consists of 33.9 km long sea dike. This dike construction led to the creation of 400 sq km of combined reclaimed land and freshwater reservoir behind the Saemangeum sea dike. The reclaimed land is used for the development of agriculture, ecology and environmental purpose and other scientific researches. The normal water level of the reservoir is at EL 1.5 m. The polder dike is a wide-based embankment with road pavement on top (EL +3.7 to +6.77 m) and consists of a sandfill core with geotubes supported by rock revetment on both sides. Sand is abundant at the site; therefore, it is used as a filler material in the core. In the design of polder dike, the rockfill berms (built up to the NWL) are used to retain the sandfill core. But in this reservoir, the rockfill revetment is substituted by the sandfill geotubes, which is also acting as a great deal of attraction.

The internal design of the dike is based on stress analysis and sand tightness. The external is based on hydraulic stability and geotechnical stability. In durability considerations, assessment risk of UV attack, abrasion impacting debris, etc., are considered during exposure. Instead of using the rockfill, geotextile tube option saves 2,30,000 tonnes of carbon dioxide emission. The construction equipment deployed in dredging, geotextile tube installation and construction of sandfill core of the polder dike includes setting barge, flat barge, tugboat, lifting crane, backhoe, dredger and anchor boat. The geotextile tubes can be alternatively used in construction to speed up the process over a certain time period when compared to the original process using rockfill berm.

4.5 Stability analysis of dikes

Sea dikes may subject to instability and settlement due to self-weight, variation in water level and weak seabed. The change in the pore pressure distribution due to change in water level affects the strength of the soil, which affects the stability of the dike (Dubey et al., 2017). Seepage failure and other failures associated with seepage are the main forms of failures in dike engineering during the flood period. Upon the infiltration of the flood, the stability problem of dike engineering is a very complicated process, which involves water movement in saturated–unsaturated states and the decrease in the strength of the unsaturated soil. Study of the dike stability mechanism under the seepage of the flood would be useful in reinforcing embankment and effective preventing damages due to flood. The effect of flood level on river dike is a typical unstable seepage process (Zhiguo e al., 2011). The drawdown rate also has a significant effect on the safety factor of dike stability. The main threats to water level rise to dike stability are indirect impacts such as accelerated erosion, scouring and water overtopping–induced soil erosion on the slopes. Slope stability of the dike becomes more complex because of seepage conditions of rainfall infiltration or the water level rise.

Computation of the factor of safety (FS) is the most critical aspect of the analysis and design of a dike. Ordinary method of slices, Bishop's method, force equilibrium method, Janbu's generalized method, Morgenstern and Price's method and Spencer's method are the different conventional limit equilibrium (LE) methods being widely used to compute the FS. For stability analysis of dikes, soil profile and water level variations are important, and there is little information available about stability analysis for a dike in an estuarine/coastal environment.

Slope stability analysis requires

a. Establishing the design and loading conditions for the embankment.

b. Performing analyses of stability for each of the following conditions. The design conditions considering for embankments are

- At the end of construction
- Steady-state seepage
- Rapid drawdown
- Earthquake loading.

The design conditions to which the slope may be subjected during its life need to be established before the stability analysis. Design cross sections of the dike must be selected on the basis of geological conditions, materials used for the construction of dike, filling materials, external forces, operational requirements, etc. The top width of the dike is chosen to satisfy the minimum requirement of having road connectivity between two ends of the dike across the lake.

LE methods have been the primary method used in determining the stability of slope for decades. The procedures are based on finding a FS for the slope. The FS shows the factor by which the shear strength must be reduced so that the reduced strength just reaches equilibrium with the shear stress. The definition of LE can be expressed in the form of Eq. (4.1).

$$\tau = s/FS \tag{4.1}$$

where s is the shear strength of the soil (give unit) and τ is the shear stress in the soil mass (give unit). Slopes can be classified as infinite slopes or finite slopes. An infinite slope represents a slope with translational failure along a single plane. The ratio of depth to failure surface to the length of the failure zone is relatively small, i.e., <10. The soil type in this failure is usually granular. Bishop's simplified method is a common method for slope stability analysis and is employed in this study. The interslice shear forces are assumed to be zero by Bishop (1955), leaving the solution overdetermined as horizontal force equilibrium will not satisfy for one slice. This method satisfies the vertical force equilibrium for each slice and the overall moment about the centre of the circular surface.

The stability analysis at the end of construction can be carried out using drained strengths in free-draining materials and undrained strengths in materials that drain slowly. This analysis is done to study the effect of pore water pressure. The side slopes become more susceptible to instability when seepage forces act along with the forces due to gravity. The shear strength will reduce with the increase in pore water pressure. Steady seepage condition is critical for the downstream slope. Due to rapid drawdown, the slopes become unstable because of the reversal in the flow direction of the water. Due to higher saturated unit weight, shear stresses are more during the rapid drawdown. But due to pore water pressure that does not dissipate quickly, the shearing resistance decreases. This condition is critical for the downstream slope. To understand the effect of seismic loading on slope stability, the analysis shall be performed considering appropriate earthquake force. Pseudostatic seismic loads can be applied in the horizontal and vertical directions by specifying the corresponding seismic coefficient. The FS values obtained for conditions with earthquake loading will be less because shear forces got generated due to earthquake forces. Besides this, the increased pore water pressure reduces the available shear strength.

4.6 Settlement of the dike

The sea dike for coastal reservoirs or any new construction on the seabed will add load to the underlying soils, thereby leading to the settlement of the soil. The total settlement has three potential components: (1) immediate settlement, (2) consolidation settlement (Sc) and (3) secondary compression. The settlement can be calculated by knowing the size of the dike, total load on the soil due to weight of the dike and the soil properties such as elastic modulus, Poisson's ratio, compression index, recompression index, secondary compression index, effective preconsolidation stress, coefficient of consolidation and coefficient of vertical compression. These properties of the soil can be determined in the laboratory after collecting an undisturbed soil sample from the site (from various locations of the proposed dike). The settlement should be calculated for the dike as even if the dike has an adequate overall stability FS, the performance can be adversely affected by excessive differential settlement. Settlement analyses for embankments over soft soils require the compression index parameters that can be obtained from standard one-dimensional oedometer tests of the fine-grained soils.

The primary consolidation and secondary compression continue to occur long after the dike is constructed; hence, they are the major settlement concerns for the design and construction of sea dike. The parameters for estimating the settlement below a dike are primarily the knowledge of the subsurface profile including soil types, layering, groundwater level and unit weights. The compression indexes for primary, rebound and secondary compression can be calculated from laboratory test data, correlations from index properties or results from settlement monitoring programs completed for the site or nearby sites with similar soil conditions. The geometry of the proposed dike, including the unit weight of fill materials and any long-term loads on the dike must be estimated for settlement analysis. Primary Sc occurs when the increase in load on soil results in the generation of excess pore pressures within the soil voids. Depending on the type of soil, the time to reduce the excess pore pressures to some steady-state level may be rapid (cohesionless soils) or may require long periods of time (cohesive soils). Therefore, primary Sc comprises of two components, the amount of settlement and the time for the settlement to occur. The amount of time required for cohesionless soil to settle is relatively short, typically occurring during construction, and the amount of settlement can be determined using immediate or elastic settlement theory. For more reading on the settlement of embankments and case study on settlement calculations for a sea dike, refer to NYSDOT (2014) and Parthasarathy et al., (2019).

4.7 Seismic hazard analysis

The sea dike constructed for coastal reservoirs may be vulnerable to earthquakes and subjected to earthquake loading. It is important to estimate the seismic hazard at the site to ensure additional seismic forces are considered while designing the dike. The design ground motion at a site is determined by conducting a seismic hazard analysis. The two approaches for evaluation of seismic hazard are the deterministic seismic hazard analysis (DSHA) and the probabilistic seismic hazard analysis (PSHA). DSHA is a simple methodology that uses geology and seismic history to identify earthquake sources and to interpret the strongest earthquake each source is capable of producing regardless of time. This is termed as maximum credible earthquakes (MCEs), the largest earthquakes that can reasonably be expected. The MCE is the largest earthquake that appears possible along a recognized fault under the presently known or presumed tectonic activity. DSHA calculates the hazard parameters based on the geological facts, and it is simple and transparent, whereas PSHA is based on numerical calculations and probability theory (Krinitzsky, 2003).

The deterministic seismic hazard analysis (DSHA) considers a particular earthquake scenario, either realistic or assumed one. The DSHA approach uses known seismic sources that are near the site and available historical seismic and geological data to generate distinct models of ground motion at the site. The earthquakes are assumed to occur on the source closest to the site. The deterministic seismic hazard analysis needs three input details like earthquake source, controlling earthquake at the source and an attenuation relation to evaluating the seismic hazard. A schematic diagram of the different steps involved in the DSHA methodology is given in. In DSHA, the controlling earthquake is assumed to act along the source at the shortest distance from the site. The uncertainties involved in the earthquake magnitude or location are not taken into account, and this method will give an upper bound value for the ground motion. Hence, DSHA method is adopted in the evaluation of seismic hazard for some of the critical structures like nuclear power plants, big dams, bridges, hazardous waste contaminant facilities, etc. The results obtained from the deterministic analysis can be used as a cap for the probabilistic analysis. Deterministic seismic hazard analysis (DSHA) provides the worst-case scenario earthquake, considering the magnitudes of the events from the point and linear sources. The hazard estimated from DSHA can be used for important structures like coastal reservoirs. The earthquake data can be compiled from agencies like USGS, ISC and available earthquake catalogues. These data should be homogenized and declustered to eliminate repetitive events, to produce a single-valued ground acceleration for different points within the area.

4.8 Summary and conclusions

Geotechnology has a significant role to play in recharging the groundwater by arresting the surface runoff, thereby allowing the water to penetrate the ground. The artificial techniques can be used for groundwater recharge, which can be classified as surface methods and subsurface methods to reduce flow velocity of the surface and subsurface water. Groundwater hydrology, soil/rock characteristics and permeability are the major factors governing the efficiency of groundwater recharge techniques. Surface methods enhance groundwater infiltration by providing more residence time. This chapter also discussed the concept of the subsurface dike or underground dams and its applicability in storing a large amount of water underground to ensure base flow in streams. Different materials with significant applications in groundwater recharge through geotechnology are also discussed in brief. Underground dams have proven to be a sustainable alternative for storing freshwater, and the benefits are many. It is the need of the hour to explore the feasibility of groundwater dams across watercourses and construct subsurface dikes with innovative geotechnical materials. Advanced research studies need to be undertaken to use geotechnology as an effective tool to aid the construction of subsurface dikes in different soil environments.

Stability analysis and settlement calculations for the sea dike are important before the construction of a coastal reservoir. On one side of the dike there is freshwater and the other side is seawater, both the levels which may vary from time to time, which differentiates this dike from normal river dikes. Through the stability analysis, the FS for the dike under different conditions can be calculated. The proposed dike with minimum functional dimensions should satisfy all the conditions of stability analysis such as the end of construction, steady seepage, rapid drawdown and seismic loading conditions. The FS values under earthquake loading is comparatively less as shear forces are generated due to earthquake forces. Besides this, the increased pore water pressure reduces the available shear strength. The FS is more when the water level is the same on both sides of the embankment, which indicates the benefit of coastal reservoirs over inland dams in terms of stability. Stability analysis and base failure are important in the study of slope stability of dike on soft soil in the estuarine/coastal environment. The total Sc might happen over a long period of time, and hence accurate estimation of the expected settlement using soil parameters and loading conditions will aid to ensure enough measures to control the settlement within limits.

References

Bishop, A.W., 1955. The use of the slip circle in the stability analysis of slopes. Geotechnique 5 (1), 7–17.

Chu, J., Yan, S.W., Li, W., 2012. Innovative methods for dike construction—An overview. Geotextiles and Geomembranes 30, 35–42.

Decker, B.L., 1986. World geodetic system 1984. Defense Mapping Agency Aerospace Center St Louis Afs Mo.

Dubey, R., Ghosh, A., Mitra, T., Choudhuri, B., Ghosh, S., Khatun, S., 2017. Stability analysis of river Dyke in estuarine environment. Indian Geotechnical Journal 47 (3), 349–363.

Fazal, M.A., Imaizumi, M., Ishida, S., Kawachi, T., Tsuchihara, T., 2005. Estimating groundwater recharge using the SMAR conceptual model calibrated by genetic algorithm. Journal of Hydrology 303 (1–4), 56–78.

Goldberg, V.M., Gazda, S., 1984. Hydrogeological Principles of Groundwater Protection against Pollution.

Iravanian, A., Bilsel, H., 2016. Strength characterization of sand-bentonite mixtures and the effect of cement additives. Marine Georesources and Geotechnology 34 (3), 210–218.

Ishida, S., Kotoku, M., Abe, E., Fazal, M.A., Tsuchihara, T., Imaizumi, M., 2003. Construction of Subsurface dams and their impact on the environment. RMZ Materials and Geoenvironment 50, 149–152. No:01.

Khairy, A.T., Al-Ghamdi, A.S., Gutub, S.A., 2010. Analysis and design of a deep subsurface dam. International Journal of Civil and Environmental Engineering IJCEE-IJENS 10 (03).

Kolathayar, S., Sitharam, T.G., Yang, S., 2019. Coastal reservoir strategy to enhance India's freshwater storage by impounding river flood waters: a detailed overview. Water Supply 19 (3), 703–717.

Krinitzsky, E.L., 2003. How to combine deterministic and probabilistic methods for assessing earthquake hazards. Engineering Geology 70, 157–163.

NYSDOT, 2014. NYSDOT Geotechnical Design Manual (Chapter 12).

Onder, H., Yilmaz, M., 2005. Underground dams – a tool of sustainable development and management of groundwater resources. European Water 11/12, 3545.
Onder, H., Yilmaz, M., 2005. Underground dams. European Water 11 (12), 35–45.
Parthasarathy, C.R., Sitharam, T.G., Kolathayar, S., 2019. Geotechnical considerations for the concept of coastal reservoir at Mangaluru to impound the flood waters of Netravati River. Marine Georesources and Geotechnology 37 (2), 236–244.
Raju, J.N., Reddy, T.V.K., Munirathnam, P., 2006. Subsurface dams to harvest rainwater – a case study of the Swarnamukhi river basin, Southern India. Hydrogeology Journal 14, 526–531.
Raju, N.J., Reddy, T.V.K., Muniratnam, P., Gossel, W., Wycisk, P., 2013. Managed aquifer recharge (MAR) by the construction of subsurface dams in the semi-arid regions: a case study of the Kalangi river basin, Andhra Pradesh. Journal of the Geological Society of India 82 (6), 657–665.
Ramasesha, C.S., Nandakumaran, P., Suresh, S., 2002. In: Dillon, P. (Ed.), Efficacy of Sub-surface Dikes as Groundwater Conservation Structures in Hard Rock Terrain of Tamil Nadu, India, Management of Aquifer Recharge for Sustainability, pp. 399–404.
Sharma, H.D., Kozicki, P., 1988. The Use of Synthetic Liner and/or Soil-Bentonite Liner for Groundwater Protection.
Sitharam, T.G., 2018. Sea dike construction challenges in coastal waters for storage of river flood waters: a sustainable strategy for water resource development using coastal reservoirs. Advances in Sustainable Urbanization 3 (1), 1–6.
Suseela, P., 2005. A low cost subsurface dike using bentonite clay- a success story. Madras Agricultural Journal 92 (4–6), 341–343.
Yang, S.Q., 2015. Coastal Reservoir-a technology that may dominate future water supply. Journal of Water Resource and Hydraulic Engineering 4 (4), 388–397.
Yee, T.W., Lim, L.K., Ter Harmsel, M., Choi, J.C., Hwang, S.P., 2013. Geotextile Tubes As a Replacement for Rockfill for the Polder Dike in Saemangeum, South Korea, Terra et Aqua (133), pp. 10–23.
Zhiguo, N., Jun, L., Ri, Y., Qing, R., 2011. The stability analysis of the dike slope in Bdg reservoir under the seepage of flood. Procedia Engineering 15, 5263–5267.

CHAPTER 5

From Dujiangyan to Qingcaosha

Yuliang Gu[1,2]
[1]National Engineer Research Center of Urban Water Resources (South), Shanghai, China; [2]Shanghai National Engineer Research Center of Urban Water Resources Co., Ltd., Shanghai, China

5.1 Introduction

The Dujiangyan is an ancient water conservancy project with the greatest age in the world. It is the only one that has been preserved and is still in use today. It is characterized by a three-stage water diversion system without any dam and has the capability to work harmoniously with the environment. The Qingcaosha reservoir is an important drinking water project with the largest water supply scale, serving a very large population and taking the longest planning time in China. This large drinking water project is located on tidal flats in middle of the Yangtze River estuary, dividing the Yangtze River into the North Channel and South Channel. It incorporates water diversion by pumps and sluices, avoiding saltwater and polluted water as well as storing fresh and clean water, uses natural self-purification, provides tidal energy storage and conforms to the river regime. Both the Dujiangyan and the Qingcaosha water supply systems are constructed upon innovative ideas utilizing the local conditions and taking advantage of the river regime, making them marvellous projects.

In the 1980s and 1990s, Shanghai Government placed priority on the development and utilization of Huangpu River water sources and made a series of strategic decisions to build Phase I and II Water Diversion Works in the upper reaches of the Huangpu River, successively with a water supply scale of 5 Mm^3/day, notably improving the quality of drinking water in Shanghai. However, it was difficult to build Shanghai into an international metropolis by solely relying on the Huangpu River water sources. Therefore, there was an urgent need to find and plan new drinking water sources. To this end, the Shanghai Municipal Water Authority research team had been seeking new water sources for Shanghai in the Jiangsu and Zhejiang Provinces and made a lot of selections, comparisons and demonstrations, but all seemed less than ideal. Finally, attention was cast on the Yangtze River estuary.

Sustainable Water Resource Development Using Coastal Reservoirs
ISBN 978-0-12-818002-0
https://doi.org/10.1016/B978-0-12-818002-0.00005-8

In the early 1990s, the study of a Planning Scheme of Yangtze River as Shanghai's Second Water Sources for Urban Water Supply was officially launched and in 1992 the Chenhang Biantan reservoir was built in the Yangtze River estuary, with a water supply scale of 1.3 Mm^3/day. However, according to the planning forecast, by 2020, the gap of good quality raw water in Shanghai would exceed 6 Mm^3/day, so it was necessary to promptly adjust the strategic pattern of Shanghai's water sources, as well as to plan and build a new water source with a larger capacity and better water quality.

Compared with the Huangpu River water sources, the Yangtze estuary water source has some unique advantages: (1) the amount of water the Yangtze River discharges into the sea is 100 times that of the Huangpu River; (2) water quality of trunk stream of the Yangtze estuary is better than that of the Huangpu River, generally meeting the criteria of Class II ~ III, according to the National Environmental Quality Standards for Surface Water (GB 3838-2002); (3) the amount of water intake from the Yangtze estuary accounts for only 0.07% of the runoff of the Yangtze River, whereas amount of water intake from the Huangpu River accounts for 27% of the runoff of the Huangpu River. Rich in water resources, the Yangtze River has excellent and stable water quality and is developed or utilized very little, while the Huangpu River water source is fully developed and utilized despite its smaller amount of water and slightly lower water quality. For Shanghai's long-term order and stability, the priority for the development and utilization of water resources needed a shift, and a new water source plan had to be made. For this purpose, the two rivers should be developed simultaneously, and the development and utilization of freshwater resources on a large scale in the Yangtze estuary became inevitable.

5.2 Planning

The Shanghai Qingcaosha raw water project consists of three main works and nine major projects. With a planned raw water supply capacity of over 7 Mm^3/day, the grand scheme was confronted with over 100 technical and management difficulties, among which the very first was the issue of planning. In which part of the Yangtze estuary should the superlarge freshwater reservoir be located? What is the service scope of the Qingcaosha raw water supply? What about the receiving water plants and their water supply capacities? How does the Qingcaosha raw water system coordinate with the Huangpu River upstream water diversion system? How does the

Qingcaosha raw water run across the Yangtze River and Huangpu River? How do water pipelines hundreds of kilometres long go through the dense city buildings? All these questions are needed to be solved one by one.

As for the issue of site selection during the reservoir planning, the research team was confronted by three major problems, including the intrusion of saltwater, stabilization of the river regime and protection of the water resources.

- *Saltwater intrusion*

First is the intrusion of saltwater. It determined whether the Qingcaosha project could be launched or not. In 1993, the research team established an automatic telemetry station for chloride in the Yangtze estuary. Based on a 15-year accumulation of long sequence data, 30 years of satellite remote sensing images and a 100-year estuary topographic analysis, a basic law about the impact of saltwater intrusion from the North Branch of the Yangtze estuary water sources was finally revealed. Thus, with factors like the Three Gorges Project, the South-to-North Water Diversion Project, the rise of sea level and the trend of river regime change in the North Branch of the Yangtze estuary taken into consideration and reviewed, the longest duration unsuitable for water intake was determined to be 68 days, providing a scientific basis for the decision-making of planning and building the Qingcaosha reservoir.

- *Stabilization of the river regime*

Second is the stabilization of the river regime. The Yangtze River is the third largest river in the world, and the Yangtze estuary has three bifurcations and four outlets to the sea, making it the most complex and sensitive river regime in the world. For any of the water area to be enclosed as a large reservoir in the Yangtze estuary, an improperly planned site and shape might lead to turbulence in the river regime, affecting the stability of the coastline and waterway and even impacting the ecological balance. To this end, by drawing on the successful experience of water management from the Dujiangyan Project, the research team established a planning idea of adapting to the local conditions and going with the river regime. They carried out seven sediment hydraulic models including two-dimensional mathematical models, three-dimensional mathematical models, an integral fixed-bed physical model and an integral moving-bed physical model. They also conducted back-to-back verification, finally revealing the complex characteristics of water, sediment and salinity in the estuary and the basic conditions needed to construct the reservoir in the Yangtze estuary project area. Meanwhile, the research team took full advantage of the tidal flat resources on the northwest end of Changxing Island in the Yangtze estuary and combined the regulation guidelines for the comprehensive planning and improvement of the Yangtze estuary together, as well as planning 16 km of levees on Changxing Island and the 32 km streamline levee around the Qingcaosha reservoir in the shape of a Chinese sturgeon. It ensures the maximization of the natural reservoir capacity and conduciveness to the river regime stability of North Channel and South Channel and north-south watercourses in the Yangtze estuary, achieving a multiwin situation for a comprehensive resource development and utilization in the Yangtze estuary.

Based on the two research results above, and through the repeated demonstration and overall planning, the water supply capacity of the Qingcaosha reservoir was finally determined as 7.19 Mm^3/day, the designed highest and lowest operating water level was determined as 7.0 and −1.5 m, respectively, the area of the reservoir was determined as 66 km^2 and the natural and maximum effective storage capacity during the construction period was determined as 385 and 435 Mm^3, respectively.

- *Protection of the water sources*

Third is the protection of the water sources. Shanghai is located beside the lowest reaches of the Yangtze River Basin, making its water quality easily affected by the incoming water from the river basin and susceptible to the urban drainage from Jiangsu Province and within Shanghai City. After long sequence and large-scale simultaneous hydrological (water, sediment and salinity) observations and rigorous verification, the research team decided to take advantage of the effect of tidal self-purification and the trunk stream barrier in the Yangtze estuary and choose a location upstream and to the north of Changxing Island as the primary water intake site and 20 km downstream as a supplementary water intake site. At the same time, the research team conducted a special study for the protection area of the Qingcaosha water sources and determined the core reservoir area of 79 km^2 as a first-grade protection zone with the edges around the core reservoir area of 70.8 km^2 as a second-grade protection zone. On 10 December 2009, before the Qingcaosha reservoir was officially put into use, the Shanghai Municipal People's Congress passed the Regulations for the Protection of Drinking Water Sources in Shanghai and adopted the most stringent protection measures for the Qingcaosha water sources, providing it with a legal basis.

5.3 Construction

Qingcaosha raw water project is composed of three main works, namely, the Yangtze Estuary Reservoir Project, the Yangtze Estuary Water Conveyance Tunnel Project and the Land-Based Pump Station and Pipeline Project. For the construction of the project, the team conducted researches over the key technologies of design and construction and solved a number of construction problems such as the closure of reservoir dykes, tunnelling water conveyance pipe crossing the Yangtze River and laying land-based pipelines.

- *The enclosure of reservoir dykes*

The Qingcaosha reservoir consists of two reservoir zones – the Zhongyang Sha and Qingcaosha, with the latter one covering an area of 53 km^2. They succeeded in enclosure at the first attempt, with no precedent in Shanghai or even in the world. Through statistical analysis of the historical data, the research team carried out dynamic topographic observations for the Qingcaosha reservoir area. In combination with an integral physical model and multidimensional hydrodynamic model of the Yangtze estuary, as well as the experience of success and failure in closing and reclamation in Shanghai's history, the research team also made repeated simulations, comparisons and selections for divided room and one-time enclosures and finally decided to follow the strategy of a one-time enclosure, just as a saying like 'better cut off one finger than hurt ten fingers'. Meanwhile, according to the plan for construction organization, for the beach section, conventional mud-filled bags were used for the hydraulic sandfill of the sloping dike structure; for the deep channel section, bags filled with sand were dumped to embank the sloping bank structure. To effectively solve the problem of river regime adjustment caused by large-scale and short-time enclosure in the complex water area at the diversion outlet, a new scheme for scour prevention and beach protection was creatively proposed, which combined the linked dyke type and the separated dyke type and dynamically implemented both types together.

The closure of the 800 m gap was a key to the success of the one-time enclosure of the 53 km^2 reservoir, and it remained as one of the worldwide problems. In the Yangtze estuary, there are two tides each day and two tidal cycles each month. Enormous energy is produced by the ebb and flow of the tides. Due to the repeated exchange between waterbodies of the reservoir and the estuary, huge amounts of tidal energy gather in the 800 m closure gap, where the maximum flow rate is up to 9 m/s based on simulation from the mathematical model. For the 800 m closure gap to the 53 km^2 reservoir, there was no diversion channel like at the Three Gorges Dam to divert river water, meanwhile it bears the bidirectional tidal flow in the estuary, making the closuring far more difficult than the Three Gorges Project. The 800 m main gap closure set a new record, based on the advanced concept of Dredging before Diversion as well as the method of Macha (a triangle timber framework used to prevent water erosion) applied in the ancient Dujiangyan system; the research team innovatively invented a three-layer combined bed-protective structure (superstrengthened sand-ribbed flexible geotextile mattress covered by a hinged concrete slab mattress and stone bagged in net) and rapid closure technology using a large-size steel cage filled with riprap and finally succeeded in the one-time closing of the 800 m gap.

- *Water conveyance pipe across the Yangtze River*

Second is tunnelling a water conveyance pipe across the Yangtze River. How to transport the raw water from Qingcaosha reservoir across the Yangtze River South Channel with a width of 7.2 km? The research team conducted special demonstrations on three construction technologies of steel pipe jacking, pipe sinking and shield-driven tunnelling, respectively, and chose the shield tunnel as the final construction scheme after comparison and selection.

There were three key technical problems to be highlighted in the shield tunnel construction scheme: (1) to achieve the maximum scouring depth and minimum covering of soil in South Channel of the Yangtze estuary, with a one-time jacking distance of over 10 km for the shield tunnelling machine; (2) to achieve a two-way pressure bearing for the water conveyance tunnel and (3) to achieve a low energy consumption of the Qingcaosha reservoir raw water. Through special researches like bidirectional stress tests for the full-size lining joint and three-dimensional numerical simulation, the process of water transportation by gravity flow with a maximum inner diameter of 5.84 m was determined, and parallel pipe laying and a single-layer lining in the shield tunnelling were adopted. The 7.2 km water conveyance tunnel represented the most advanced process of one-time medium-diameter shield tunnelling.

- *Construction of pump station and underground pipelines*

Third is the extra-large water supply pump stations and the under-crossing of land-based underground pipelines. The Wuhaogou Pump Station is an extra-large comprehensive water transportation pump station, with a water supply capacity of 7.08 Mm^3/day and functioning in water conservancy and municipal service. Due to the limited planned land, the pump station covers an area of only 93 m^2. For safety, low power consumption and land-saving of the pump station, the interactive verification of integral CFD 3D numerical simulation and physical model tests, as well as an optimized system integration technology, was applied in the urban water supply pump station for the first time. Innovation was made in the process arrangement of the extra-large multisystem pump station, and the objective of steady flow and balanced flow distribution was achieved. The extra-large water supply pump station was built upon a 12,000 m^2 underground foundation pit at a depth of 22 m with 24 large water pumps (total power exceeding 80,000 kW) and the maximum lift head of 70 m.

The land-based pipeline project includes the Changxing Island Branch, Lingqiao Branch, Yanqiao Branch, Jinhai Branch and Nanhui Branch, with a length of over 200 km. The research team used the multilevel series pressure boost technology and a one-way pressure supplement tower to establish a new energy-saving and safe method for the design of hydraulic pressure boosting and water hammer control in the large-diameter and long-distance water transportation pipeline system, minimizing its impact on the aboveground urban buildings and safety of the raw water dispatch (Fig. 5.1).

5.4 Operation

As the Qingcaosha raw water project is a huge systematic project, top-level design for its safe operation in the planning stage is a must. Regarding the running of the project, the research team carried out studies on key technologies concerning the operation and solved a series of operational problems such as eutrophication, energy consumption of water diversion and delivery and water quality early warning and dispatch.

- *Eutrophication in the reservoir*

First is the eutrophication within the 66 km^2 reservoir area. Qingcaosha reservoir is located in the lower reaches of the Yangtze River, so its waterbody contains abundant pollutants and nutrients. On the basis of ensuring the reservoir's function of avoiding salty water and storing fresh water, and according to the laws of hydrological and water quality change in the Yangtze estuary, the research team took full advantage of the tidal self-purification effect of the Yangtze estuary and creatively designed the function of avoiding polluted water and storing clean water at the intakes of the reservoir.

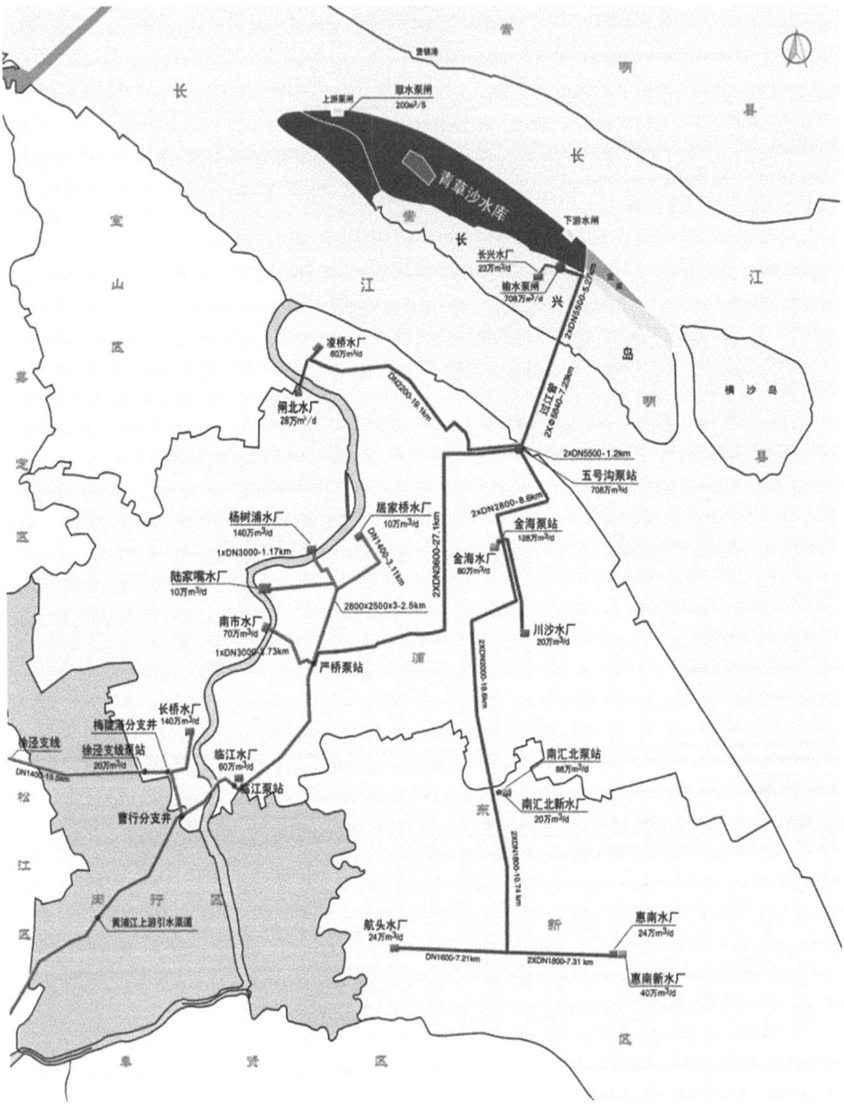

Figure 5.1 System diagram of the Shanghai Qingcaosha raw water project.

The Yangtze estuary waterbody is characterized by high turbidity. As a type of natural coagulant, sediment has a good effect of purification in the waterbody, together with the slow flow of the reservoir water along a length of 22 km, making the water quality significantly improved. However, Shanghai is located in a subtropical region where the temperature of air and water are both high, and the waterbody in the Yangtze estuary contains rich nutrients, making a stagnant or microflowing waterbody an easy target for eutrophication or even algal outbreaks. Therefore, in the planning stage, the research team conducted a special study on the retention time of the reservoir water and concluded that a water exchange period of 20 days is needed in the reservoir during summer. Based on this calculation, the coordinated operation of upstream and downstream sluices, with the capability of bidirectional flow, was installed in the Qingcaosha reservoir to increase the water mobility and reduce retention time of the waterbody. Meanwhile, in the planning and running stages, the team continued to research of the integration and application of precontrol technology for water eutrophication in the Qingcaosha reservoir area and comprehensively improved the water quality in the Qingcaosha reservoir by applying biological, physical and chemical methods.

- *Energy consumption of water diversion and delivery*

Second is the energy consumption of water diversion and delivery. The Qingcaosha reservoir has a water supply capacity of 7.19 Mm^3/day. Considering the supply and exchange of water, its water diversion capacity exceeds 10 Mm^3/day. Under the precondition of ensuring the reservoir's water quality, and to solve the problem of high energy consumption for the water pump stations in traditional water conservancy projects, the research team adopted an alternative operation mode with water diversion from the inlet and outlet water sluices as the major water movement mode and water from the pumps as a supplement. (a) In the period of no saltwater intrusion in the Yangtze estuary, water diversion using the tidal effect of the Yangtze River through the sluices is applied as far as possible, which not only saves energy consumption in the pump stations but also makes full use of the tidal energy storage in the Yangtze River to raise the reservoir's water level. (b) Before the saltwater intrusion period in the Yangtze estuary, water diversion by pumps during high tide levels is applied as far as possible.

As the length of Qingcaosha water transportation pipelines exceeds 200 km and in response to the problem of high energy consumption of extra-large water delivery pumps and sluices for the long-distance water delivery system, the research team made a number of strategic decisions: (a) The raw water is transported by gravity flow across the river through the water delivery tunnel with an inner diameter of 5.84 m, and the reservoir water delivery Wuhaogou Pump Station with a capacity of 7.08 Mm^3/day is cancelled. (b) In the pump station a variable frequency control technology is applied for all the 24 large water pumps which are synergistically dispatched with flow pressure from the pressure regulating tanks and receiving water plants along the water delivery pipeline.

- *Early warning and dispatch of the water quality*

Third is an early warning and dispatch of the water quality. To achieve the refined dispatch of quality water from the Qingcaosha reservoir, as well as coping with any accidental water pollution in the upper reaches of the Yangtze River, the research team has strategically built a multilevel early warning and monitoring system for the hydrology and water quality of the Yangtze River and Qingcaosha reservoir. (a) Cooperate with the hydrological bureaus in the upper reaches of Yangtze River to share the hydrological and water quality data of the Yangtze River. (b) Monitor the Yangtze estuary chloride telemetry system composed of 15 monitoring sites. (c) Design a network for the early warning, forecast and dispatch of the reservoir's water quality, composed of hydrographic forecast, online chloride testing, biological monitoring, floating automatic water quality monitoring station, fixed automatic water quality monitoring station and water quality monitoring ship. (d) Build a three-level water quality analysis laboratory. (e) Establish the Qingcaosha water quality testing zone and the Qingcaosha operational intelligent dispatch system in the Yangtze estuary.

5.5 Conclusions

The Qingcaosha water sources raw water project has eight advantages: adequate freshwater resources, stable and good water quality, a large availability of water quantity, feasible protection of the water sources, conducive to the river regime of the South Channel, huge water supply potential, strong antirisk capability and significant scale effect. It has shown notable comprehensive benefits since it was officially put into work in June 2011.

- *Significant improvement of water quality in the water sources*

For water quality in the Qingcaosha reservoir, 21 basic parameters meet the standard of Class I and 1 meets the standard of Class II. Of the total 24 basic parameters, only total phosphorus (lake and reservoir) and total nitrogen do not meet the standard continuously. The 5 supplementary parameters and 80 specific parameters all meet the standard, according to the National Environmental Quality Standards for Surface Water (GB3838-2002).

- *Significant improvement of water quality in water plants*

Due to the improvement of raw water quality, the amount of coagulant used in the water plants receiving Qingcaosha raw water is reduced by about 50%, and 106 parameters of the effluent fully meet the National Standards for Drinking Water Quality (GB5749-2006), with most below the detection limit of the standard.

- *Significant social benefit*

Water supply capacity of the project exceeds 7.19 Mm^3/day, accounting for over 50% of the total raw water supply in the city, with 13 receiving water plants. The supplying area covers all seven administrative districts including the Pudong New District and part of five administrative districts such as the Chongming District, benefiting a population of over 13 million.

Acknowledgements

The author greatly acknowledges the contribution made by Weizhi Wang and Chenmin Ruan who have assisted in the English translation for the paper.

References

National Environmental Quality Standards for Surface Water (GB3838-2002). Ministry of Ecology and Environment, China.

National Standards for Drinking Water Quality (GB5749-2006). Ministry of Health, China.

CHAPTER 6

Insights into the design and development of Shanghai coastal reservoirs

Jian Zhong Yuan, Cai e Wu
Shanghai Investigation Design & Research Institute Co., Ltd (SIDRI), Shanghai, China

6.1 Development of Shanghai coastal reservoirs

Shanghai is a mega city with water shortage induced by water pollution. Shanghai city water supply is mainly from the Huangpu River, and the rest of the water was taken from adjacent rivers and groundwater before the 1990s. The Huangpu River is located downstream of the Taihu Basin. Freshwater quality and water supply can be easily affected by accidents and pollutants in the basin. Other rivers in Shanghai are scattered, and their overall water quality is poor. Water quality in these rivers generally cannot reach the national surface water standard for a centralized drinking water source. Therefore, these rivers cannot be used as the main drinking water sources of Shanghai.

At the same time, the groundwater yield is low. To effectively control land subsidence, further control and reduction in the rate of groundwater exploitation is needed in Shanghai. With the development of Shanghai's economy and the growth of urban population, the demand for freshwater is on the increase. The number of residents in Shanghai was 24.2 million in 2016. The water supply capacity of the city reached 11.52 million cubic metres per day, and the total water supply was 3.2 billion cubic metres per year. In addition, there is a shift in the emphasis placed on the demand for water from quantity to quality. The lack of good water quality will have an impact on the sustainable development of Shanghai. Therefore, finding new water sources of adequate quantity and high quality for urban water supply has become a pressing problem.

The search for drinking water sources in Shanghai goes back for more than 100 years and can be divided into three stages. The first stage was the period before the 1980s when the city mainly took water directly from the lower reach of the Huangpu and Suzhou Rivers. The second stage began in

Sustainable Water Resource Development Using Coastal Reservoirs
ISBN 978-0-12-818002-0
https://doi.org/10.1016/B978-0-12-818002-0.00006-X

1987, when the city started taking water from the upper reach of the Huangpu River. The third stage is represented by the Qingcaosha reservoir, a large-scale exploitation and utilization of freshwater resources in the Yangtze River estuary.

The Yangtze estuary has abundant freshwater resources of good and stable water quality. The water resources in the Yangtze estuary account for 98.8% of the total water resources in Shanghai. The water quality meets the requirements of a centralized drinking water source and has relative high self-purification ability. It is better than surface water in other areas of Shanghai. The freshwater resources of the Yangtze estuary have great potential for development and utilization. Therefore, the development of freshwater in the Yangtze estuary can not only satisfy the incremental freshwater needs of Shanghai but also improve the quality of the Shanghai water supply. The drinking water supply of Shanghai will mainly depend on the freshwater in the Yangtze estuary. But the direct development in the Yangtze estuary is threatened by saltwater intrusion at the intakes. So, a reservoir with appropriate storage capacity must be constructed to store freshwater during periods of low salinity and high quality.

Shanghai has three coastal reservoirs used for drinking water sources in the Yangtze estuary. They are the Chenhang reservoir, the Qingcaosha reservoir and the Dongfengxisha reservoir (Fig. 6.1). The characteristics of the three Shanghai coastal reservoirs can be seen in Table 6.1.

6.2 Layout of the Qingcaosha reservoir

The Qingcaosha reservoir is located on the northwest side of Changxi Island that forms the diversion between the South Channel and the North Channel of the Yangtze estuary. Fig. 6.2 shows the general layout of the Qingcaosha reservoir. The main function of the reservoir is to store freshwater during periods of low salinity and to use it in the dry season. The reservoir can provide enough freshwater for the water receiving area even under the circumstances of the longest continuous unfavourable water intake time of 68 days during saltwater intrusion. The morphological characteristic of the reservoir is long and narrow. It is about 22 km long and 3 km wide. A combined pump and sluice upstream water intake and downstream sluice operation scheme and general layout are proposed on the basis of a comprehensive study of the tide characteristics, saltwater intrusion, fluvial processes, flow pattern and water quality evolution in the reservoir. The water intake (including upper sluice and pump) is arranged at

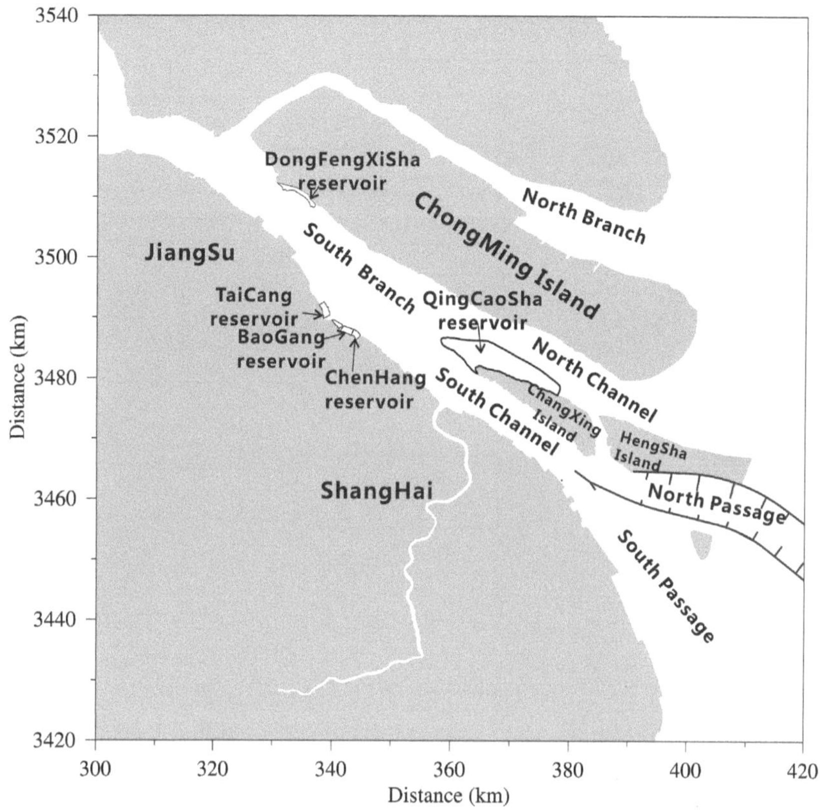

Figure 6.1 The location of the Shanghai water sources in the Yangtze estuary.

Table 6.1 The characteristics of the three Shanghai coastal reservoirs.

Reservoir	Area (km^2)	Total capacity	Daily water supply	Continuous supply time	Operating data
Chenhang	1.336	9.5 Mm^3	1.3 Mm^3/d	7 d	June 1992
Qingcaosha	66.15	0.527 Gm^3	7.19 Mm^3/d	68 d	June 2011
Dongfengxisha	3.74	9.762 Mm^3	0.215 Mm^3/d	26 d	January 2014

the upper end of the reservoir. The water supply system and lower sluice are arranged in the downstream section of the south and north sides of the reservoir, respectively. This arrangement can make full use of the tide and take water at higher tidal levels, reducing operating energy consumption. At the same time, the operation scheme that takes water from upstream and drains water from downstream parts of the reservoir can control the water retention time in the reservoir.

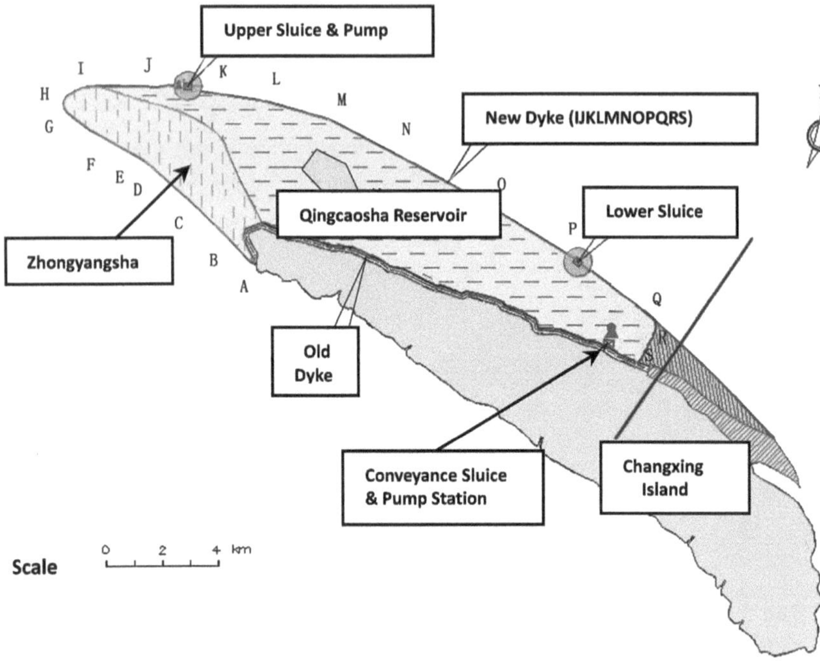

Figure 6.2 The general layout of the Qingcaosha reservoir.

The total length of the dyke is 48.4 km, and the newly constructed dyke (IJKLMNOPQRS) is 22 km. Total reservoir area is 66.15 km^2. The upper gate net width is 70.0 m (5 m × 14.0 m), and bottom elevation is −1.50 m. The upper pumping station capacity is 200 m^3/s. The lower gate net width is 20.0 m (5.0 + 10.0 + 5.0 m) and bottom elevation is also −1.50 m. Maximum high water level in the reservoir is 7.0 m (Wusong Datum). The dead water level is −1.5 m. Total reservoir capacity is 0.527 Gm^3, and the design effective capacity is 0.438 Gm^3. Design daily water supply is 7.19 Mm^3/d. Freshwater is extracted from the reservoir using a gravity water conveyance method. The net width of the water conveyance gate well is 24.0 m, and bottom elevation is −4.0 m. Design water conveyance scale of pump for Changxing Island is 0.11 Mm^3/d.

6.3 The key concepts in the design

6.3.1 General consideration of natural and construction conditions

The natural and construction conditions include meteorology, hydrology, hydrodynamics, runoff, tidal and wave conditions, fluvial processes, geology, water quality, construction material, site condition and adjacent relations. The design parameters were based on a study of the aforementioned conditions. The main design parameters included design tidal level, wave, saltwater intrusion duration, geological parameters, control velocity and tidal level process. The design flood control (tidal) standard for the Qingcaosha reservoir is the 100-year high tidal level plus the same frequency wind and waves. Water supply guarantee rate is no less than 97%. Allowable surf amount under design wave condition is 0.05 $m^3/(s.m)$. Seismic fortification intensity is 7°. Continuous unfavourable water intake time is 68 days. The tidal type of maximum tidal range over 10 years in the nonflood season was adopted for calculating the maximum flow speed in damming through the closure gap and the design protection velocity for the closure gap. The flow velocity in the closing gap process was controlled by no more than 6.0 m/s. The design maximum protection velocity of the closure gap (at the junction of the platform and the slope) is 7.5 m/s, and the velocity on other parts on the platform surface is 6.0 m/s.

6.3.2 Site selection and arrangement of dyke line

There are three bifurcations and four outlets into the sea in the Yangtze estuary, namely the North Branch, the North Channel, the North Passage and the South Passage. The fluvial processes and saltwater intrusion are complex. The following factors needed to be considered in site selection for the coastal reservoir: (1) the water quality meets the required standard; (2) there is enough freshwater; (3) scouring and silting is relatively small; (4) there are no signification effects on the regime changes, navigation channel and important facilities. The site selection needs to consider the impacts of the project implementation on the estuarine fluvial evolution; on the other hand, it needs to consider whether the regime changes will bring security impacts on the project itself and the corresponding engineering measures that need to be proposed. The site of the reservoir and dyke alignment should be selected in the areas where the riverbed/seabed is relatively stable and no serious scouring/silting occurs.

The impact assessment had been conducted on the basis of some technical means, e.g. fluvial evolution analysis, mathematical models and fixed and movable bed physical model tests. The characteristics and trend of the fluvial evolution at the reach of the Qingcaosha reservoir had been systematically studied on the basis of research and comprehensive summarization of the existing results, combining the analysis of the flow, sediment and underwater topographic data. It was revealed that the regime of the Qingcaosha site was relatively stable before 2007. It was at the critical point of erosion and moving down of the diversion area between the South Channel and the North Channel of the Yangtze estuary. It was a good opportunity to build the Qingcaosha reservoir. There were no obvious adverse effects on flood control and drainage, navigation channel, fluvial processes, existing engineering and facilities in the Yangtze estuary.

6.3.3 Law of saltwater intrusion and reservoir scale

The law of saltwater intrusion has a direct impact on an estuarine reservoir. It plays a key role in the site selection, the layout of hydraulic structures and the operation of the reservoir. Saltwater intrusion from the North Branch of the Yangtze estuary increases the complexity of the saltwater intrusion law in the Yangtze estuary. When and how to take freshwater depends on the interactions between runoff and tide. The longest continuous unfavourable water intake time has a decisive influence on the reservoir scale.

The scale of the Qingcaosha reservoir is determined by the scale of freshwater supply and maximum number of consecutive unsuitable water intake days caused by saltwater intrusion. Therefore, a study on the law of saltwater intrusion was one of the key factors.

Saltwater intrusion is affected by the combined effects of runoff, tide, plane form, wind, Coriolis force and circulation off the shelf. It has a characteristic of remarkable temporal and spatial variation. Upstream runoff and tide are the main influencing factors of saltwater intrusion in the Yangtze estuary. Saltwater intrusion from the North Branch to the South Branch increases the complexity of the law of saltwater intrusion in the Yangtze estuary. The salinity of the Qingcaosha water is affected by both the saltwater intrusion from the North Branch and the saltwater intrusion from the sea. Correlation analysis and mathematical model analysis show that the longest continuous unfavourable water intake time is 68 days during saltwater intrusion period around the Qingcaosha reservoir during the 1978–79 design representative typical year.

6.3.4 Intake method and operation scheduling

There are three methods of reservoir intake, namely pumping, sluice and combined pumping and sluice. The changes of water quality, saltwater intrusion and river regime evolution directly influence the reservoir water intake reliability. The rational choice of the position of the water intake and the method of water intake has a direct impact on water supply safety, scale of water intake project, reservoir operation cost and getting high-quality freshwater. Tidal characteristics and the law of saltwater intrusion should be studied. Pump and gate location and operation should be reasonably arranged. An appropriate scheduling operation can reduce operating costs and save energy.

Qingcaosha reservoir has plenty of freshwater from May to October. The prestorage water in the reservoir is needed when the intake water salinity is not suitable for water intake during the saltwater intrusion period from November to April. The highest water level in the Qingcaosha reservoir is 7.00 m. The mean annual high tide level adjacent to the Qingcaosha reservoir is 3.35 m and the mean annual lower tide level is 0.92 m. There is plenty of freshwater in the flood season, and the reservoir is maintained at a constant water level. Tidal power is used to bring freshwater into the reservoir through the sluice. The water level of the reservoir is filled as high as possible by the sluice, and then the pumping station is used to pump freshwater into the reservoir during the dry season. So a combined sluice and pump method is used in the dry season. This method takes into account the hydrological characteristics and the saltwater intrusion characteristics around the Qingcaosha reservoir. Because sluice intake can be used for most of the time in a year, this intake method reduces energy consumption.

The water body in the reservoir is prone to eutrophication because of higher nitrogen and phosphorus contents during the flood season. So the sluice in the lower part of the reservoir is arranged, and it is used to drain water from the reservoir. Through the upstream and downstream linkage operation mode, the mobility of water in the reservoir is increased (Fig. 6.3), and the residence time of water in the reservoir is reduced. The ability of reservoir to resist eutrophication is also enhanced. The residence time of water in the Qingcaosha reservoir is controlled as 15–20 days at present.

The water level in the reservoir is maintained as high as possible for energy saving and is generally at 3.3 m. Reducing the water level may be needed to increasing the mobility of water for water quality control.

To meet the demand of freshwater supply during the saltwater intrusion period, the water level in the reservoir has to be raised through using both the sluice and pump. Reservoir water level is generally 6.2 m during the saltwater intrusion period. Pumping more freshwater into the reservoir before the advent of a serious saltwater intrusion is based on prediction.

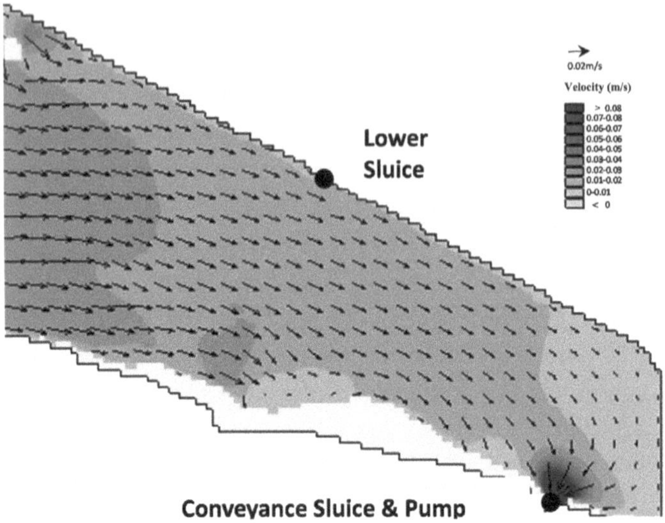

Figure 6.3 Flow field in the lower part of the Qingcaosha reservoir.

6.3.5 Water quality protection and reservoir eutrophication prevention

The reservoir has the characteristics of a shallow lake. The flow pattern in reservoir also has an influence on water quality. The reservoir shape, layout of pump and gate and scheduling operation need to be fully considered to make the water flow more smoothly and to maintain a reasonable water residence time as an algae control measure.

The water flow in the reservoir slowed down after the completion of the reservoir, and the residence time of water in the reservoir became longer. There is stagnant area in some local areas. There is a greater risk of eutrophication, even the risk of algal blooms, when higher contents of nitrogen and phosphorus are present in the reservoir water. This easily leads to deterioration of water quality in the reservoir and may have a certain impact on the quality of freshwater.

The Qingcaosha reservoir is a large water supply reservoir. The only way to prevent eutrophication was to optimize the shape and morphology of the reservoir and to select the appropriate layout and optimize the reservoir operation scheme to increase the flow of water and establish a sound ecological system and other measures to maintain and improve the water quality of the reservoir.

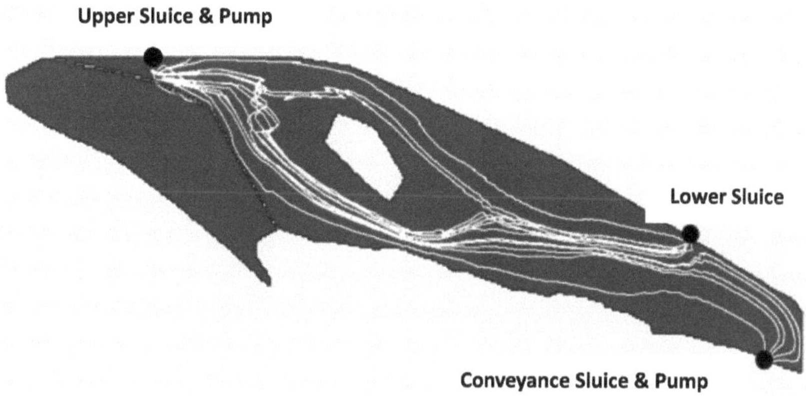

Figure 6.4 A schematic diagram representing the trajectory of water points.

A long and narrow reservoir in morphology was formed, which is more than 20 km long with an average width of 3 km. This feature on the one hand can ensure that the effect on the coastal regime is reduced to a minimum. On the other hand, it can ensure that the water body has a certain amount of mobility. The general layout of water intake from upper section and drainage from lower section is adopted by the reasonable layout of the water intake, drainage and water distribution. Through the linkage of the operation of the upstream and downstream facilities, the mobility of water in the reservoir is increased (Fig. 6.4), and the residence time of water in the reservoir is reduced. Dredging to make the water flow more smoothly can also reduce the risk of local algal outbreaks. The implementation of ecological construction in stagnation waters near north dyke can reduce the risk of algal blooms. Raising or lowering the operational water level and increasing water flow according to the characteristics of algae growth is also used to prevent algal blooms.

6.3.6 Hydraulic fill dam in soft soil foundation

6.3.6.1 General consideration

The dykes around coastal reservoirs are generally constructed in a seawater environment where the water depth is generally around 10 m. Therefore, conditions such as deep water, fast flows and high waves are unfavourable for construction. Furthermore, the foundation soil is often unfavourable to stabilize the dyke due to its soft and weak nature and severe settlement potential. Because the long dykes for many such reservoirs are built into open water, there is no land for activities supporting the construction of the dykes.

Also the construction cannot incur large-scale changes (erosion and siltation) of the seabed. In addition, the most suitable dyke type should be chosen to save investment because of the lack of stone in the estuarine area. So the following points may be considered: (1) sequence of dike construction, to avoid any massive scouring and silting of the movable seabed; (2) dam type and structural type of deep-water dike; (3) seepage proofing of the dyke.

The natural conditions at the Qingcaosha site are a bidirectional current, mobile bed and soft soil foundation. The average tidal range and maximum tidal range are 2.43 and 4.40 m, respectively. The average tidal level is 2.12 m. The dike at the eastern part of the Qingcaosha reservoir was built crossing a deep passage of a flood tide channel where the elevation was −10.5 m. In such cases, specific, plastic drainage plates can be used for foundation treatment, and large and high strength geotextile bags with hydraulic filling can be placed to form a sloping dyke.

6.3.6.2 Structural type of deep-water dyke

The part of the dyke crossing the thalweg accounts for one-fifth of the newly built dyke for the reservoir, including the 2.2-km-long eastern section where the maximum water depth was up to 15 m, the seabed surface lies at an elevation range of −5.0 to −10.5 m and dyke height ranges from 13 to 19 m. The foundation surface is a 3- to 5-m-thick silt layer overlying a 10-m-thick highly compressive sludge. The hydrological features are deep water, fast flow, reversed and variable tide. From the perspectives of soft soil foundation treatment and hydraulic filling in deep water for dyke construction, a new dyke structure and associated construction technology were formulated. Specifically, plastic drainage plates would be used for foundation treatment, and large and high strength geotextile bags with hydraulic filling were used to form a dual-prism sloppy dyke structure (Wu et al., 2017).

6.3.6.3 Seepage proofing of a hydraulic filled dyke

The dyke foundation consists of complicated silt and sludge beds in conditions of flood and ebb flow. Field tests, generalized physical modelling of dyke sections in a flume, and in situ testing of adjacent or similar dyke's permeability were used to unveil the permeability and permeable damage mechanism of the new sedimentary soil and hydraulic filling dyke in the Yangtze estuary. The following laws of permeability were gained: (1) a single-layer geotechnical bag structure is close to the original sand sample in permeability; (2) a multilayer geotechnical tubular bag structure generally has a two to four times reduction of the filling body permeability, which is mainly caused by connecting seepage through the geotechnical tubular bags; and (3) permeable damages all occur in connecting seepage of geotechnical tubular bags, and the critical gradient value of connecting seepage is 0.1.

Through the field test of impermeable wall and the improvement and optimization of construction control requirements, the problems such as triaxial mixing piles penetrating geotechnical bags and pile shaping are resoundingly solved. An impermeable wall of triaxial cement mixing piles was successfully applied to the hydraulic filling dyke that totals 30 km in length. As for the closure section of the dyke, a new type of impermeable wall was built by assembling one row of high-pressure jet grouting piles between two rows of triaxial mixing piles (Wu et al., 2017).

6.3.7 Closure gap setting, protection and closure

The coastal reservoirs in Shanghai were formed by building an encircled dyke. Therefore, the greatest concern and most risky point during the construction period were the arrangement and protection of the closure gap, as well as the closure design. Because of the effect of the reversing tide at the closure gap, the periodic tidal variation limited the continuous and available working time. Due to the shortage of stone material and land riprap, the soft and weak soil foundation can be easily washed away, and it is difficult to protect. Any scouring is extremely difficult to control.

6.3.7.1 Hydraulic parameters of closure gap

The study was mainly performed through 2D and 3D hydrodynamic mathematical models, hydrodynamic physical model tests for the closure gap, box cage wave tank tests and cross-section tests in a flume. All these means were mutually combined and complementary. Upon the study results, an evaluation and comparison was conducted using several construction factors, and finally the hydraulic parameters of the closure gap were determined. To suit the actual conditions during construction and based on the tidal and meteorological changes, a short-duration hydraulic forecasting technology of gap closure was developed and used in practice. The design flow speed at the closure gap by vertical and plain plugs was 9.5 and 7.5 m/s, respectively. According to technical and economic comparisons, the closure method was finally determined to be a plain plug.

6.3.7.2 Protection of closure gap

Through comparison of several schemes, setting a closure gap (800 m wide) at the deep flood channel in the eastern dyke section was finalized. Within the closure gap, the maximum design flow speed was 7.5 m/s. The armour face of the closure gap adopted a multilayer composite protection structure; the layers are, from the bottom up, large-sized and high-strength geotechnical cloth bags filled with sand (sand quilt), 1300 g/m^2 super-strong geotechnical cloth soft drainage body of sand ribs, soft drainage body of chained concrete blocks, and 60 tons of string bagged stones or 30 tons of concrete blocks (Lu and Wu, 2013).

6.3.7.3 Method of damming

For the damming of the closure gap, a plain plug with steel-caged riprap was used, and the design maximum flow speed for damming the closure gap was 7.5 m/s (real-time measured speed was 5.15 m/s).

Three problems needed to be solved in the design of the steel cage, i.e., (1) the stability of cage during initial placement; (2) contacting status of riprap in the cage with the string bagged stones on subgrade; and (3) evenness of the cage placed on the subgrade of string bagged stones. Through the testing and analysis of the stability of cages under the effects of various flow conditions, the volume of each steel cage was finalized as about 1000 m^3, and a permeable supporting skeleton made by shaped steel was used.

Prior to damming, taking the advantage of small neap tides, the steel cages were placed from the two sides of the 800-m-wide closure gap towards the middle, and simultaneously, ripraps were dumped at the bottom and both sides of the cages to add weight against the flood and ebb flow; thus, a permeable skeleton of the damming structure was formed. To complete the damming, the duration of next small neap tide was picked, and a fleet of ships was set up to dump riprap into the cages.

The plain plug method with large-sized steel cages filled with riprap was successfully used in damming the Qingcaosha reservoir. The application of the method created the record for damming a 49.8 km^2 water area without separating working places and the discharge of 150 Mm^3 through the closure gap in one tide duration.

6.4 Conclusion and prospects

So far, the Shanghai coastal reservoirs have performed well, and the water quality of the reservoirs meets the requirements of the design standards. The key concepts adopted in the coastal reservoir design have been proved to be correct and reasonable. It can provide technical reference for similar coastal reservoir studies, design and construction. However, some areas are still worthy of study.

(a) Because the Qingcaosha reservoir is in the diversion waters between the South Channel and the North Channel of the Yangtze estuary, the fluvial process is extremely complex. The hydrodynamic processes at the reach of the Qingcaosha reservoir and fluvial development trends, especially in the waters near the western end, the upstream section at the north side of the Qingcaosha reservoir, should continue to be monitored. Beach protection measures should be carried out if necessary.

(b) The residence time of water in the Qingcaosha reservoir was planned to be 15—20 days. The eutrophication risk in this reservoir is relatively high because nitrogen, phosphorus and other nutrients are high in the Yangtze estuary. A study about how to increase the water mobility to shorten the residence time to 7—10 days and further reduce the risk of an algal bloom would be useful.

(c) It is recommended to continue dredging and to optimize the reservoir underwater topography to enable the flow conditions to meet the requirements for ecological restoration both inside and outside the reservoir embankment.

(d) The capacity of the Chenhang reservoir is small. It cannot meet the water demands because of the growth of the population. Research is needed on how to expand the capacity and improve the water supply guarantee rate of the Chenhang reservoir.

(e) Water quality at the intake of the Chenhang reservoirs is easily affected by irregular drainage from the Liuhe River of Jiangsu province because the Chenhang reservoir is located downstream from the Liuhe River. Optimizing the reservoir operation scheme is still needed.

(f) Other potential water sources in the Yangtze estuary are still worth studying to meet the water demands of Shanghai in 2040, when the population is predicted to reach 25 million people.

Acknowledgements

Shanghai Investigation Design & Research Institute Co. Ltd (SIDRI) was the main designer of the Qingcaosha reservoir. The authors would like to thank all participants for their involvement in the Qingcaosha reservoir design.

References

Lu, Z., Wu, C., 2013. Key technology and practice of dyke engineering design of the Qingcaosha reservoir. Journal of Yangtze River 44 (21), 28–31 (In Chinese with an English abstract).

Wu, C., Yuan, J., Wei, Q., 2017. The key design concepts and technologies of theQingcaosha reservoir in Yangtze Estuary. In: Proceedings of the 37th IAHR World Congress. Kuala Lumpur, Malaysia, pp. 6239–6248. August 13-18, 2017. Paper No. 08-41-1600.

CHAPTER 7

Preliminary feasibility study of coastal reservoirs for Australia

Shu-Qing Yang, Muttucumaru Sivakumar
School of Civil, Mining and Environmental Engineering, University of Wollongong, Wollongong, NSW, Australia

7.1 Introduction

According to the World Bank (2019), at present, 2.2 billion people lack reliable access to safe drinking water, and 1.8 billion people will be living in regions or countries with absolute water scarcity. The world population of 7.7 billion in 2019 is estimated to be increasing by 1 billion every 12 years. More than 50% of the world population now lives in cities. An estimated 150 million people now lives in cities with continually recurring water shortages, and this number is expected to increase rapidly to 1 billion by 2050 (World Bank, 2019). The epitome of the global water crisis can be seen in Australia, the driest inhabited continent in the world with 25 million people in 2019 spread over 7.7 million km^2, where desert makes up about 18% of the mainland area of Australia. The size of the Murray–Darling River basin is similar to that of the Ganges River in India (about 1 million km^2), but during the Millennium Drought (2000–09) in Australia, the Ganges River flow was 200 times higher than that in Murray–Darling River. The long-term average rainfall across the whole country is 457 mm/year (i.e., 3.52×10^6 GL/year); only 11% of it becomes the mean annual run-off from Australian catchments, i.e., 390×10^3 GL/year (www.bom.gov.au).

To seal the gap in water supply and demand, many possible solutions have been proposed in the world, and Australia has tested these ideas and proposals, including the following:

7.1.1 More inland dams

In the 20th century, the dam industry played a leading role in securing water supply. Australia often claims to have the highest storage per capita in

Sustainable Water Resource Development Using Coastal Reservoirs
ISBN 978-0-12-818002-0
https://doi.org/10.1016/B978-0-12-818002-0.00007-1

the world. In the 21st century, it has become more difficult to construct large dams because almost all sites with ideal hydrosocial and geotechnical combinations have already been dammed. Many negative impacts on the riverine ecosystems caused by the inland dams have been found (Yang, 2003, 2004), and the public firmly oppose new dams due to the negative social and environmental impacts, e.g. the proposed new Tillegra dam in NSW. Consequently, governments, research communities and the public have shifted to other alternatives.

Table 7.1 shows the year of the last major dam constructed for Australian capital cities, and from it, one can clearly see that in the future it is almost impossible to construct any new dams for these capital cities. During the Millennium Drought (2000–09), the available water was only 18%–33.8% of the current dam capacity along the wetter east coast, i.e. for Brisbane, Sydney and Melbourne. It is inferred that other capital cities such as Adelaide and Perth had no water above their dead storage during this period.

After the 2015 national census, the Australian Bureau of Statistics published its prediction of populations in Sydney and Melbourne, and the results are shown in Fig. 7.1. It can be seen that, compared with 4 million people in 2005, these cities will double their populations to 8 million in 2050. It is a natural question to ask where the drinking water sources are going to come from for the additional 4 million people ? Column 6 in Table 7.1 shows that during the past 30–40 years, all capital cities have had no new dams, and it is predictable that in the next 31 years (from 2019 to 50), no new dams will be built for these capital cities' water supply. Therefore, state and local governments are looking for new solutions for the future water supply.

7.1.2 Wastewater reuse

Wastewater reuse and recycling in cities have been adopted in many countries including Japan, the United States, Israel, Namibia, Singapore and Australia. However, the costs of treatment including the energy intensity to convert wastewater to drinking water standards are high. Another important aspect is the "yuck" factor that many Australians do not wish to drink treated wastewater (sewage). However, in Namibia, which is among the most arid countries in the world, where its capital, Windhoek, receives an annual rainfall of 370 mm (Lahnsteiner and Lempert, 2007), wastewater reuse for drinking water has been practiced for decades.

Table 7.1 Australian capital cities' last dam constructed and lowest % of its capacity during the Millennium Drought (2000–09).

Capital city	Last dam	Storage capacity (GL)	Dam's catchment (km^2)	Year completed	Age of the last dam (years)	Lowest water in 2000–09
Sydney	Tallowa	85.5	5750	1976	43	33.8% in 2007
Melbourne	Thomson Reservoir	1123	487	1983	36	26% in 2009
Brisbane	Wivenhoe	1165	7020	1985	34	18% in 2007
Adelaide	Little Para	20.8	83	1977	42	≈ 0
Perth	Ten Mile Brook	1.5	4	1994	25	≈ 0
Darwin	Copperfield	0.5	0.68	1987	32	n.a.
Hobart	Craigbourne	12.5	241	1986	32	n.a.

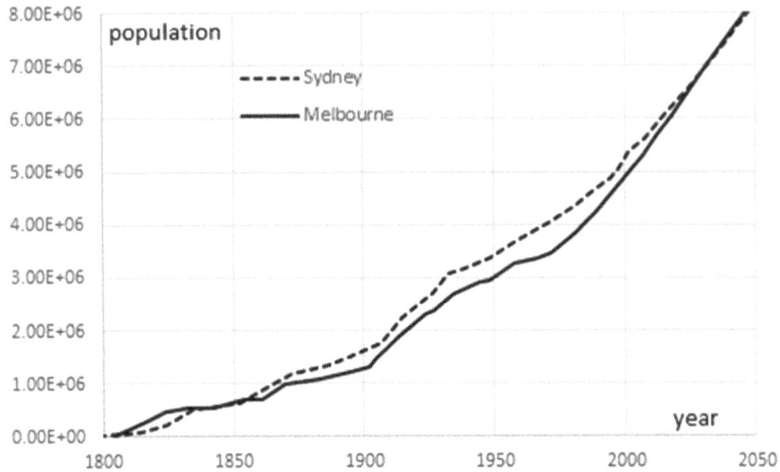

Figure 7.1 Australian Bureau of Statistics' prediction of population growth in Sydney and Melbourne, both will double in size from 2005 to 50.

7.1.3 Desalination plants

This has been seen as a promising way to supplement climate independent water supplies especially during droughts. There are now 92 desalination plants in Australia. Every capital city in Australia has constructed and is operating one or more desalination plant. However, their environmental footprint is large, capital and operational costs are high and compared with inland dams and reservoirs and their design life is also short.

More countries have realized the importance of the development of new stored water technologies, and the water resources development should be extended from upstream mountainous areas to downstream areas, and into the sea. Globally, current water consumption is 2.3×10^6 GL/year and will be 3.57×10^6 GL/year in 2050, with this huge water demand representing only 5%–6% of global run-off lost to the sea (i.e., 40×10^6 GL/year). In 2013–14, Australia used 23.5×10^3 GL/year of water for agriculture, industry and households (Bureau of Meteorology, 2015). Compared with the annual run-off to the sea which is approximately 390×10^3 GL/year, the used water is only 5.9% of total run-off, similar to other areas in the world. In the next 100 years, only 1%–2% of additional run-off is needed to sustain the growth in water demand.

7.2 The first- and second-generation coastal reservoirs

Yang (2015) claimed that the Australian water crisis is not caused by its water shortage, rather by storage or reservoir shortage; it is time to construct coastal reservoirs to harvest floodwater lost to the sea. Many attempts have been made in the world to construct coastal reservoirs, and most existing coastal reservoirs can be classified as the first-generation coastal reservoirs, which stores water on the continent side relative to the coastline. Such in-stream water bodies generally have shortcomings such as poor water quality, high construction costs, significant environmental impacts and inappropriate size. The second-generation coastal reservoir stores water on the seaward side relative to the coastline; the off-stream water body can well overcome the aforementioned problems. The first-generation CRs have the following common features:

(1) Solid and very expensive dams or barrages are used to separate the seawater and freshwater. In other words, the construction method of inland dams has been extended to the coastal reservoirs, which leads to a costly barrage. To reduce its construction cost, the dam is often constructed in the places where the dam or barrage has the shortest length.

(2) The first-generation coastal reservoirs have the potential to collect every single drop of run-off from its catchment and also collect the pollutants from the catchment, such as industrial chemicals, agricultural pesticides and fertilizers. As a result, these coastal reservoirs may become a wastewater reservoir like what occurred in the Sihwa Lake in South Korea. With increase in population economic conditions, the production and consumption pattern of communities also increase. This will more likely to add further wastewater (domestic, industrial and agricultural) discharges and associated pollutants into the river systems that can end up in the coastal reservoir.

(3) Except for coastal reservoirs in Shanghai, China, all existing reservoirs' barrage alignment is perpendicular to the flow direction. This barrage alignment perpendicular to river flow, similar to inland dams, significantly changes the river flow and causes many environmental and ecological problems, e.g. navigation and passage of migratory fish, etc. It is expected that in the future, no coastal reservoirs that totally cut off the river flow will be approved by any government or water authorities.

The second-generation coastal reservoirs (Yang et al., 2001) aim to solve the problems in the first-generation coastal reservoir and achieve the following purposes:

(1) The construction cost for coastal reservoir is reduced significantly.

(2) The quality of stored water in coastal reservoirs is better than the water quality in the existing inland dams.

(3) The impacts of coastal reservoirs on society, environment and ecosystem are acceptable to all communities.

The second-generation coastal reservoir is different from the existing coastal reservoir in terms of design philosophy, barrier design method, operational methods and the capital/running costs.

The new design philosophy is that floodwater lost to the sea is also a water resource worth developing. Nature is a giant desalination plant that uses the solar energy to evaporate seawater, and its desalination process includes evaporation, condensation and precipitation. The process of natural desalination is very effective for energy saving. A number of problems associated with engineered seawater desalination processes have been identified before. Using a second-generation coastal reservoir, it is possible to capture and store floodwater for water supply purposes before floodwater entering the sea, and hence, one does not have to go through desalination processes.

As the sea is the biggest reservoir in the world, the coastal reservoir can be constructed at any place near the sea shore. The primary barrier is located, at least in part, in a body of seawater. In addition, the reservoir may have a secondary barrier located within a storage zone defined by the primary barrier, with the secondary barrier creating a buffer zone between the primary barrier and the secondary barrier so that freshwater contaminated with seawater will occur in the buffer zone.

As for barrier design methods, almost all existing coastal reservoirs in the world have short barrage lengths because the designers try to reduce the construction cost. This implies that the designers assume that the barrages are very expensive and the benefit of a very long barrage may not be able to cover its cost. Consequently, the barrages are always perpendicular to the flow direction, which cuts off ship navigation and fish migration, causing significant environmental and social impacts. As these reservoirs cannot adjust their size to match the water demand, very long hydraulic residence time often results in deterioration of water quality. One example is the five barrages at the outlets of Lake Alexandrina, South Australia, which was constructed by Engineers Australia in the 1930s to change this brackish water body into a freshwater lake by collecting water from Murray River. The high evaporation over its very large surface of shallow water, coupled with low river flow in the Millennium Drought (2000–09), led to salinity in the lake being higher than seawater salinity. This highlights the failure of conventional barrage designs that are applicable to first-generation coastal reservoirs.

In the second-generation coastal reservoir's design, the improvements include the following. The barrier of the coastal reservoir may not be as sturdy as the concrete inland dams. Its primary barrier can be a solid barrier with a height sufficient to withstand normal tidal surges and wave action along the coast. Concrete dams, breakwaters, seawalls, caissons, earth/rock dykes or very large sandbags can be used for its construction. For example, a 170-km-long dyke was constructed to protect the navigation channel in Yangtze estuary in 2002 using prefabricated caissons and sandbags/geo-synthetic tubes, where the water depth ranged from 5.0 to 8.5 m and wave height was 3.32–5.90 m (Yan and Chu, 2010). The total cost including dredging of 0.3 billion m^3 of soil in the channel was $15 billion Yuan RMB or US$2.5 billion, giving US$15 million/km of cost for a solid barrier. The water level difference on both sides of inland dams can be up to 200 m, which requires very rigid concrete to resist the huge pressure force. However, when the water level difference on both sides of a coastal reservoirs' barrier is usually less than 10 m, dykes that are used in coastal structures such as harbors and land reclamation projects can be used.

Seepage from inland dams could be a big problem, but for a coastal reservoir, this is insignificant because the pressure difference is small, and most of the time, the reservoir's water level is higher than the sea level. Seawater intrusion can be further prevented by introducing a soft dam that is similar to a silt protector. The curtain is suspended from the boom placed on water surface and extended to the seabed to separate freshwater and brackish water. The buffer zone between the solid and soft barriers traps seawater from the top and bottom by surges and seepage of groundwater, respectively, to minimize potential contamination, so fresh and high-quality water is protected further by the soft barrier.

Rubber dams can be used as the intake/tidal gates. A rubber dam has been used in the Singapore Marina coastal reservoir. It becomes a dam when inflated with water or air, and the dam height is lowered when it is deflated (Chu et al., 2012).

Therefore, compared with first-generation coastal reservoirs, the construction cost of second-generation reservoirs could be much cheaper.

Its barrage length should not be a main concern in the coastal reservoirs' design, but its appropriate size or ratio of water volume to its surface area is the main concern. Stagnant water bodies and evaporation loss should be minimized. In other words, for an inland dam design, its water level or dam height is the critical parameter to determine. For a coastal reservoir design, the water level is almost constant (i.e., the sea level), and surface area is more important. Shallow water should be avoided in the reservoir, since evaporation caused the failure of first-generation coastal reservoirs like Lake Alexandrina.

7.2.1 Operational methods

The most important feature of a second-generation coastal reservoir is to store good quality and to discharge polluted run-off out of the reservoir without mixing. In contrast, existing coastal reservoirs such as Singapore Marina Barrage generally receive polluted water (i.e., first flush) but discharge clean water to the sea (after sedimentation). By using the SPP (separation, prevention and protection) strategy (Yang and Liu, 2010), water quality could be improved by separating freshwater from seawater using a barrier, clean water from polluted river water using intake regulation, and only good quality water is allowed to enter the reservoir. The collected freshwater needs to be protected against polluted river water and external pollution. Seawater intrusion into the stored freshwater should be prevented. The SPP strategy is effective in collecting good-quality water, because the intake gates will be closed when poor-quality water or first flush flows appear in the river mouth, allowing the polluted water to flow to the sea via a bypass channel. The quality of reservoir water could be further improved if a wetland pretreatment is allowed.

Generally, good-quality water appears during flood seasons except its first flush that should not be allowed to enter the coastal reservoir, as the first flush transports the majority of contaminants such as suspended sediments, total nitrogen (TN), total phosphorus (TP) and other nutrients and heavy metals. These pollutants generally come from nonpoint sources over a land surface, particularly after a long spell of dry weather. Normally, it is not necessary to collect all run-off, as globally only additional 1%–2% of run-off is enough for human's water demand. It is suggested that multiple pretreatment wetlands can be applied to reduce nutrients such as TN and TP to ensure the reservoir's water quality.

The most fundamental difference between the first- and second-generation coastal reservoirs is the bypass channel and selection of run-off based on its quality. The first-generation coastal reservoirs have no bypass channel to discharge polluted water, and poor-quality water is allowed to enter the reservoir and to pollute the high-quality water inside the reservoir, like the Sihwa coastal reservoir in South Korea, which was abandoned for water supply after its construction (Bae et al., 2010). But if bypass channels were used in the reservoir, the Sihwa Lake could restore its water quality and function as a water source.

7.2.2 Operational problems and solutions

The biggest challenge for a coastal reservoir is not water quantity, but contamination by seawater from groundwater seepage or overtopping, or land-based wastewater. This problem is solvable using multiple barriers and bypass channels. The river flow is not disrupted, and fish can swim back to the river's upper course for breeding. This new type of coastal reservoir can minimize the environmental impacts, and a win—win solution can be achieved for water supply, the ecosystem and affordable costs as summarized in Table 7.2.

To find the possible environmental impacts of coastal reservoirs, the author keyed in 'environmental impacts of Zuiderzee' in Web of Science, one of the largest research databases, and only one paper by Lammens et al. (2007) appeared on 18 February 2017. Their research reveals that the impacts of the Zuiderzee reservoir include 'damming and fixing the water table prevented the development of emergent vegetation and caused steep

Table 7.2 Cost comparison among different methods of water supply for per m^3 of water.

	Inland dams	Desalination plant	Wastewater reuse	Coastal Reservoir
Design life span (year)	100	20	20	100
Capital cost (A$/ m^3/year)	0.23	1.0	1.0	0.01
Running cost (A$/ m^3/year)	0.40	7.1	6.5	0.40
Total (A$/$m^3$/year)	0.63	8.1	7.5	0.41

water-land gradients'. Other impacts are 'high nutrient loads, which cause phytoplankton blooms, the disappearance of aquatic macrophytes and intensive fishery'. However, when keyed in 'environmental impacts of Three Gorge Dam (TGD)', the largest inland dam in the world completed in 2007 in the same database, 769 journal papers appeared. When 'environmental impacts of desalination' are keyed in the same database, 3697 journal papers appear. The 85 years since Zuiderzee construction is long enough for researchers to observe the coastal reservoir's environmental impacts, but it receives the least environmental concerns relative to inland dams and desalination plants.

7.2.3 Capital and running costs

The proposed Traveston Crossing dam in SE Queensland could supply 70 GL/year with AU$1.6 billion for the dam construction and 76 km^2 of land to be flooded (Turner et al., 2007). The designed life span of the concrete dam and its steel structures is about 100 years; thus the capital cost is $0.23/$m^3$/year. Its running cost is much cheaper compared with Sydney's desalination plant that costs AU$535M for its state of hibernation or A$5.9/$m^3$ of manpower cost. The designed life span of desalination plants is about 20 years. This water desalination plant needs a power consumption of 4 kWh/m^3 (Stamatov and Stamatov, 2010); thus, the energy cost is about $0.3 × 4 = AU$1.2/m^3, so the total running cost is AU$7.1/$m^3$. The capital cost for 90 GL/year is AU$1.8 billion; for 20 years life span, it gives A$1.0/$m^3$/year.

The Western Corridor Project invested AU$2.6 billion for a water supply capacity of 130 GL/year; thus, its capital cost is AU$1.0/$m^3$/year for its 20-year designed life span. Its manpower cost is similar to the desalination plant, but its required energy is less, and A$0.6/$m^3$ is assumed. This gives the wastewater recycling method a running cost of AU$5.9 + 0.6 = AU$6.5/m^3.

The Qingcaosha coastal reservoir in Shanghai spent AU$3.7 billion for the construction cost including a 45-km-long dike, pumping system with the capacity of 200 m^3/s or 2600 GL/year, 114-km-long pipeline system and a 7.2 km underground tunnel (about 6 m in diameter) and two sluice gates with widths of 70 and 20 m (see Chapter 6 for details). The designed life span of the coastal reservoir is also 100 years; thus, its capital cost is AU$0.01/$m^3$/year, and its running cost is very low as the government charges the residence's tap water at AU$0.4/$m^3$ only. In this study, inland dams' running cost is assumed to be AU\$0.4/m^3, same as the coastal reservoir. Table 7.3 provides the comparative costs among different methods, where the total cost means the capital cost plus its running cost. It can be seen that inland dam's total cost is 1.5 times higher than coastal reservoir's total cost. Desalination and wastewater recycling are approximately 19 times higher.

7.3 Is Australian water crisis caused by storage shortage?

The previous analysis shows that Australia and the rest of the world use only 5%–6% of its run-off; therefore, it is not that Australia is running out of water, it is that water is running out of Australia. A natural question is why not increase dam numbers and storage capacity to solve the water crisis. Fig. 7.2 shows the Australia's dam construction over the past 100 years (Larsen et al., 2014). It is clear that the rapid rise in dam construction from the 1960s to the 1980s petered out in the 1990s, and after 2000, there were almost no new large new dams constructed, which is because more and more people have realized the significance of a dam's negative impacts on the ecosystem, environment and the society. Additionally, no more suitable dam sites are left after the intensive construction of dams in the 20th century.

In water resources engineering, the worst enemy of sustainable use of reservoirs is sedimentation (USBR, 2006). The total sediment yield in the world is estimated to be 13.5 × 10^9 tonnes/year or 150 tonnes/km^2, and about 25% of this is transported into the seas and oceans, whereas 75% is trapped, retained and stored in the lakes, reservoirs and river systems (Batuca and Jordaan, 2000). Consequently, the silting process is reducing the storage capacity of the world's reservoirs by more than 1% per year. UN experts warned that a fifth of the current storage capacity of reservoirs worldwide or 1500 km^3 will be gradually lost over the coming decades as a result of sedimentation. Furthermore, it has been shown that global warming may increase the severity of storms and droughts, which accelerate the natural erosion rates that feed reservoirs. As the climate warms, there is more water vapour in the atmosphere causing more extreme storm events. The life span of world reservoirs depends on the sedimentation rate, which is reducing the storage capacity of the world's reservoirs by more than 1%–1.5% per year (White, 2001).

A concrete dam's design life span is about 100–200 years. In this study, the average life span of a dam is assumed to be 150 years, the total dam

Table 7.3 Basic information for the proposed interbasin water diversion schemes in Fig. 7.15.

Water source	Annual flow (GL/year)	Distance to another river basin, X (km)	Highest elevation of diversion route (m)	Elevation at source (m)	Water destination/ elevation (m)	Water lifting height H (m)
Morgan on Murray River	75	359	460	17	Whyalla/60	267 (class I)
Mannum on Murray River	140	60	480	15	Millbrook dam, Adelaide/300	322 (class II)
Mundaring Weir in Perth	8.4	530	450	130	Kalgoorlie/411.5	612 (class III)
Kimberley in Western Australia	200	1900	670	30	Perth/300	1457 (infeasible)
Burdekin dam in Burdekin River	7200	180	400	158	Thomson River in Lake Eyre Division/400	354 (class II)
Fairbairn dam in Fitzroy River	3000	110	590	200	Warrego/ Condamime –Culgoa Rivers/ 590	458 (class II)
Wivenhoe dam in Brisbane River	300	70	484	100	Condamine –Culgoa River/ 484	428 (class II)
Warragamba dam in Hawkesbury River	600	50	800	141	Arthur River/800	690 (class III)

Thomson dam in Thomson River	400	25	1000	435	Goulburn River/1000	580 (class II, tunnel)
Lake Argyle in Ord River	3940	200	400	87	Mackay, Wiso/400	438 (class II)
Coastal reservoir at Bega River	570	50	1500	0	Murrumbidgee River/1500	1531 (infeasible)
Coastal reservoir at Clarence River	4920	80	1000	0	Border River/1000	1050 (class III)
Wellington Lake	3600	100	1000	0	Upper Murray River/1000	1062 (class III)
Coastal Reservoir at Carpentaria	6530 (Nicholson—Norman River)	250	350	0	Georgia River/350	506 (class II)
Coastal Reservoir at Fitzroy, WA	6000	180	186	0	Great Sandy desert/186	298 (class I)

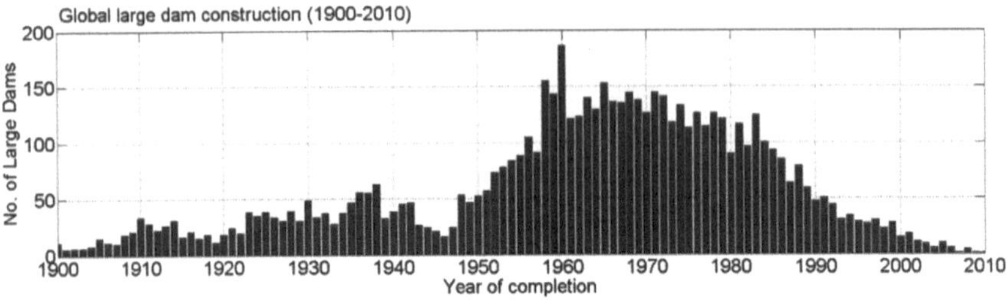

Figure 7.2 The number of large dams constructed in the world by year (Larsen et al., 2014).

number and reservoir storage capacity can be calculated, and the results are shown in Fig. 7.3, which clearly indicates that by 2150, almost all reservoirs will disappear and the water infrastructures will lose their function. Fig. 7.3 also shows that in the human history, the period from 1990s to 2050 possesses the highest number of active large dams and the highest effective storage capacity (6200 km^3). Fig. 7.3 clearly shows that our water crisis is intensified by sedimentation and aging concrete steel structures.

Australia had a good record in dam construction during the 1960s to the 1990s as shown in Fig. 7.4A, which is based on the database of the Australian National Committee on Large Dams (ANCOLD). Among the total dam numbers, more than half of the dams were designed for other purposes, rather than water supply, for instance irrigation, flood control and hydropower. The number of dams for water supply is shown in Fig. 7.4A. The main problem for Australian dams is that, different from the United

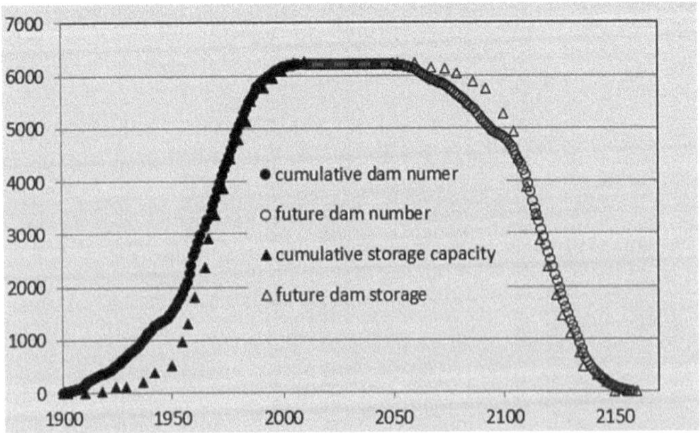

Figure 7.3 The cumulative large dam number and storage capacity (GL) in the world for the period of 1900–2160 based on the assumption of 150 years dam life span.

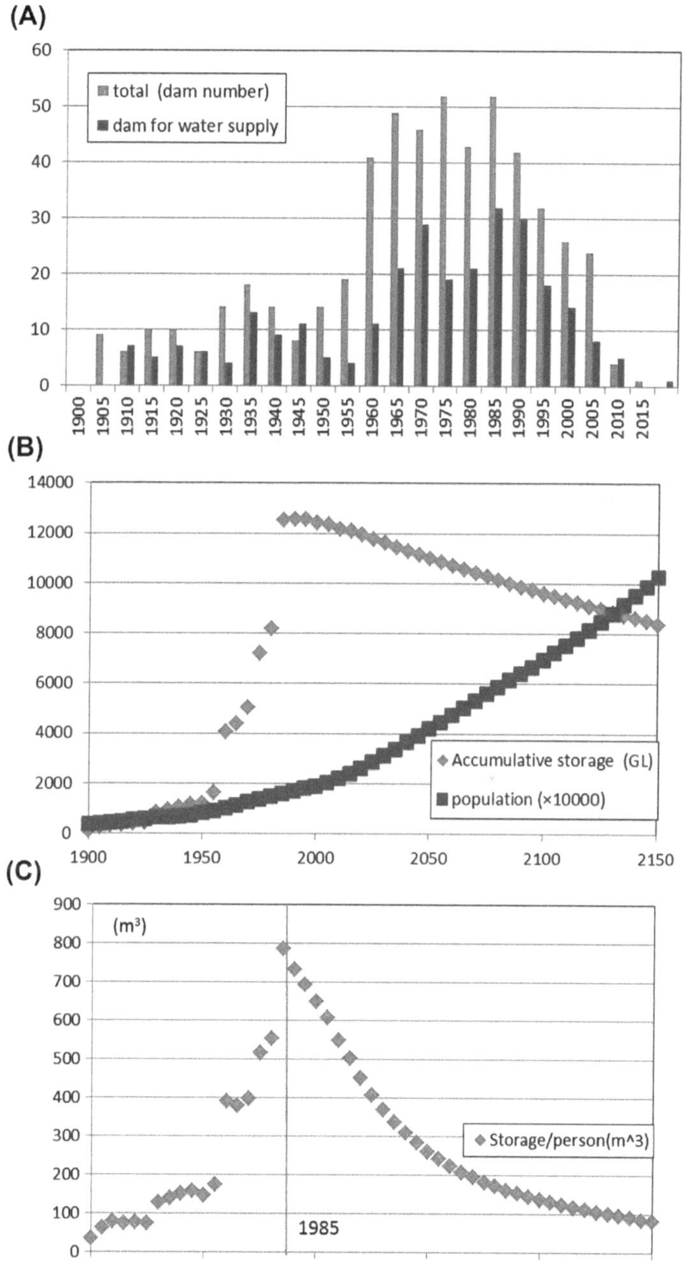

Figure 7.4 (A) The number of Australian dams constructed; (B) population growth and total storage capacity; and (C) storage capacity per capita from 1900 to 2150.

States and China, there are limited high mountains and deep valleys. The flat land makes Australian reservoirs shallow with a large water surface. Consequently, for a given reservoir storage, Australian dams generate much higher evaporation loss relative to other countries' deeper dams in cooler climates. This is why the benefit of dam construction in Australia often cannot cover its cost. For example, the Ord River Scheme in Western Australia cost AU$613 million between 1958 and 1991, but the benefits were just AU$102 million in the same period (Larsen et al., 2014).

Australia has the highest population growth among the major developed countries as shown in Fig. 7.4B, which is based on the data provided by Australian Bureau of Statistics. White (2001) estimated that the annual sedimentation rate is approximately 0.27% of reservoir's storage, approximately one-fourth of the world's average. In other words, the expected half-life of Australian dams is 185 years. This means that in the future, if no new dams are developed, the storage volume will decrease significantly as shown in Fig. 7.4B.

Australia is very proud of its achievement—its 500 large dams yield the highest per capita surface water storage capacity in the world. This was needed because Australian rivers are characterized by low discharges and the highest variability (Finlayson and McMahon, 1988; Puckridge et al., 1998). However, this statement may not be true as shown in Fig. 7.4C, because Australia had its highest storage capacity per capita (800 m^3/capita) in 1985, but this value has decreased continuously due to the population growth and sedimentation. Fig. 7.4C also reveals the hidden threat that at least one-third of Australia's storage capacity per capita has been lost since 1985. We cannot continue to claim that Australia has the highest storage capacity per capita in the world. Very soon, by 2050, our storage capacity per capita will be slightly more than that during World War II as shown in Fig. 7.4C. Obviously, we need more reservoirs to overcome the next drought, like the Millennium Drought during 2000–09. This is why in the world many countries have seen a shift in the philosophical paradigm of dam development from mountainous regions to the downstream (coastal) storage as claimed by Malaysia.

7.4 New water solution for Australian capital cities

To solve Australian water crisis, coastal reservoirs for the capital cities have been proposed in Figs 7.5–7.14. Its conceptual design is given as follows: red lines (grey lines in printed version) represent solid dams or shoreline of

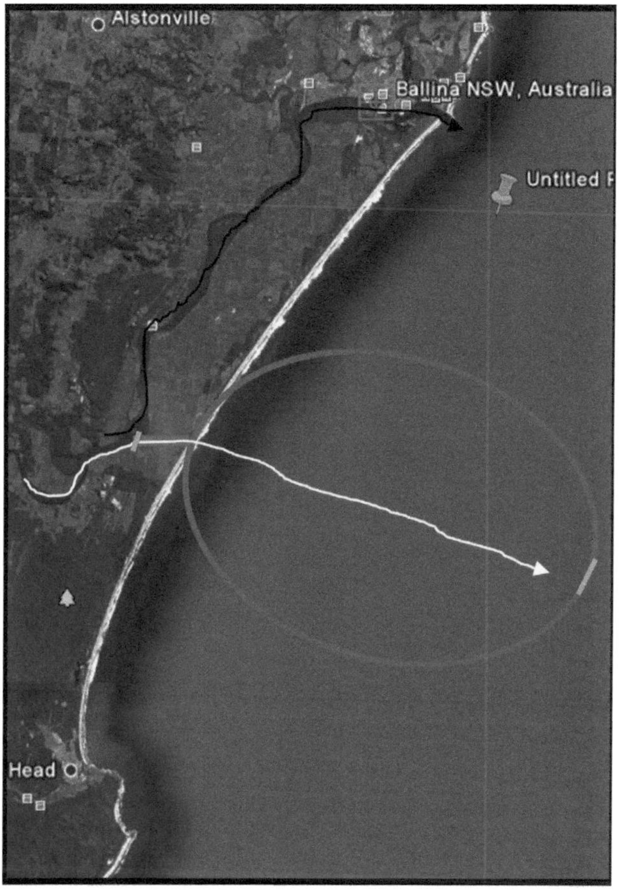

Figure 7.5 Proposed coastal reservoir at Richmond River (100 km from Gold Coast)

freshwater reservoirs, the artificial canal is represented by an orange thick line and the flow path of high-quality river water is represented by a white line with an arrow. The hydraulic structures such as intake/tidal gates are represented by short yellow lines (light grey lines in printed version). The flow path of poor-quality water or the by-pass channel is represented by a black line with arrow. In Australia, all capital cities except Hobart have faced water crisis, which can be seen from the emergence of desalination plants. Hence, it is necessary to analyze whether the water crisis is caused by water shortage or storage shortage and whether coastal reservoirs can quench Australia's thirst.

Southeast Queensland covers about 22,000 km^2 extending 240 km from the Gold Coast to the Sunshine Coast and 140 km from the coastline

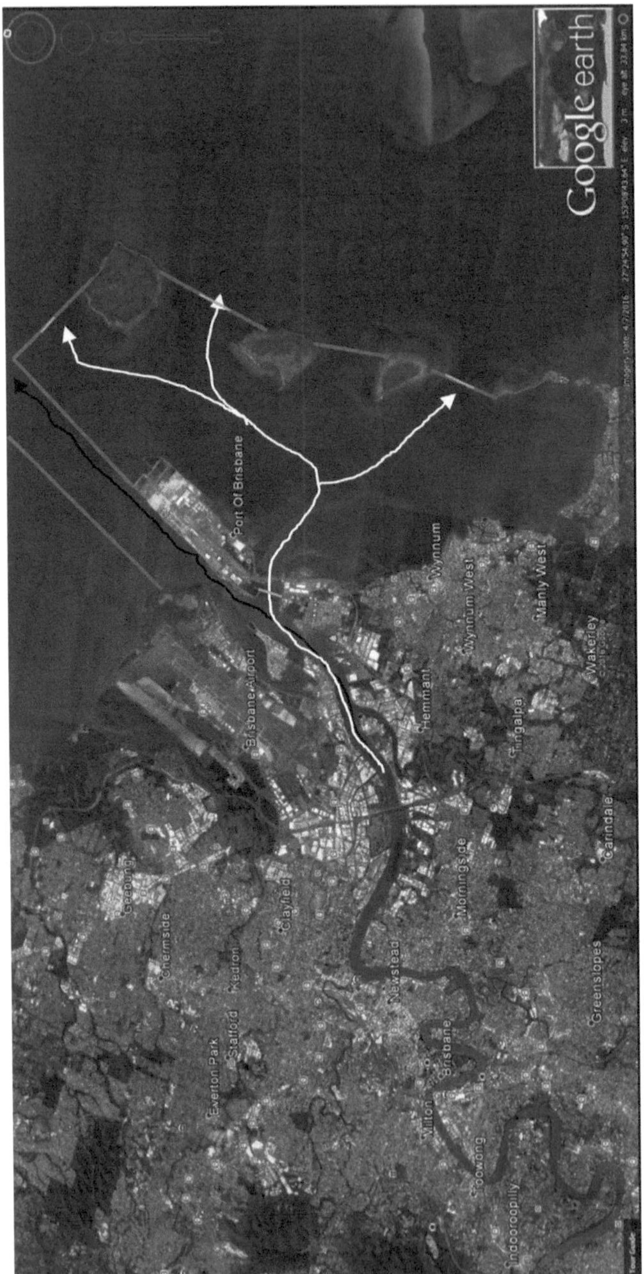

Figure 7.6 Proposed coastal reservoir at Brisbane River mouth.

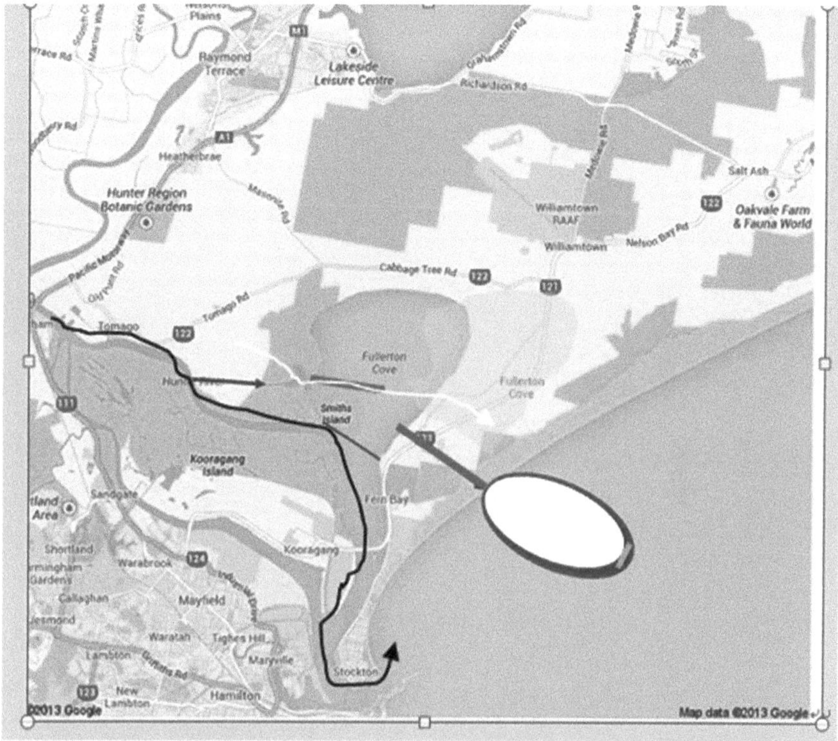

Figure 7.7 Proposed coastal reservoir for Newcastle at Hunter River mouth.

to the Great Dividing Range in the west. This region's average rainfall is around 1200 mm/year; the total run-off from Richmond River in NSW, Brisbane River and Mary River is 7100 GL/year, but its water supply in 2015 was only 300 GL (Seqwater, 2015), the available water is 24 times more than the water demand. To solve this region's water crisis, the following coastal reservoirs have been proposed by Yang (2016), as shown in Fig. 7.5 and Fig. 7.6, at the mouths of the Richmond River and the Brisbane River, respectively. The annual run-off of Richmond River is 3300 GL/year from its 6900 km^2 catchment. The annual run-off of Brisbane River is about 2500 GL/year. In Fig. 7.6, the total length of dyke is 16.3 km, and the water surface area of coastal reservoir is 66 km^2.

NSW's total population in 2015 was 7.7 million, with Sydney 4.9 million or 64%, and Newcastle—Maitland 0.4 million or 5.6%. During the Millennium Drought (2000—09), water in Warragamba dam dropped to its lowest recorded level of 33.8% on 9—10 February 2007. Warragamba

Figure 7.8 Proposed coastal reservoir at the Hawkesbury River mouth by 2.9 km barrage.

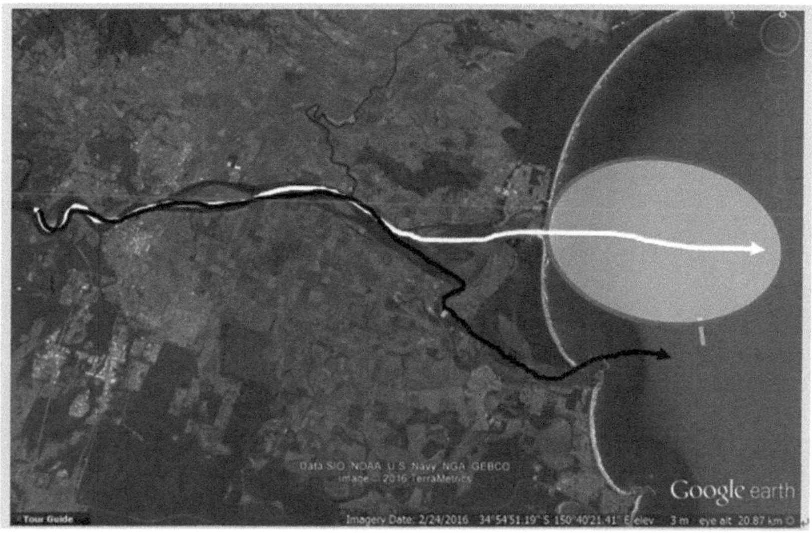

Figure 7.9 Proposed coastal reservoir at the mouth of Shoalhaven River.

Figure 7.10 A proposed coastal reservoir to develop the wetland pretreated water from Yarra River using pipelines or underground tunnel.

dam's catchment is only 9000 km^2, but the catchment areas of the Hunter River, Hawkesbury River and Shoalhaven River are 22,000, 21,600 and 7000 km^2, respectively. If the coastal reservoirs proposed in Figs 7.7–7.9 are applied, the total catchment is as high as 50,600 km^2, which is 5.6 times larger than Warragamba dam's catchment. It is certain that these regions'

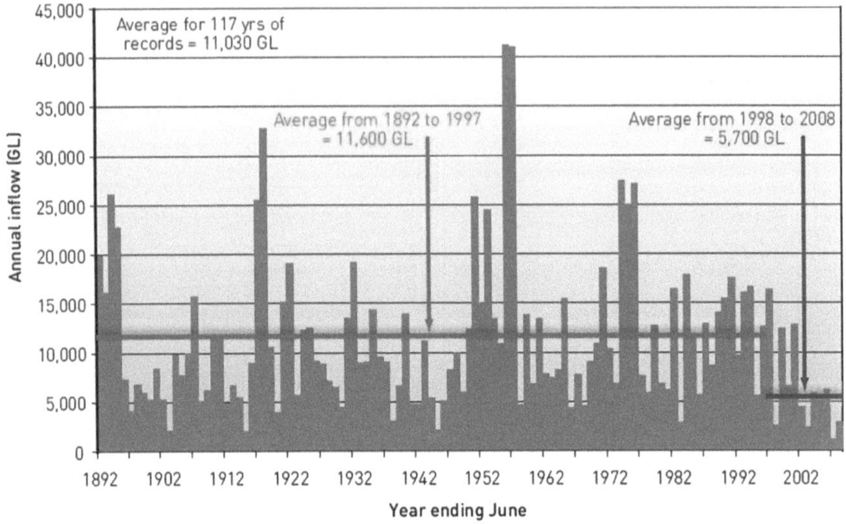

Figure 7.11 Murray–Darling River run-off from 1892 to 2008 to Lake Alexandrina.

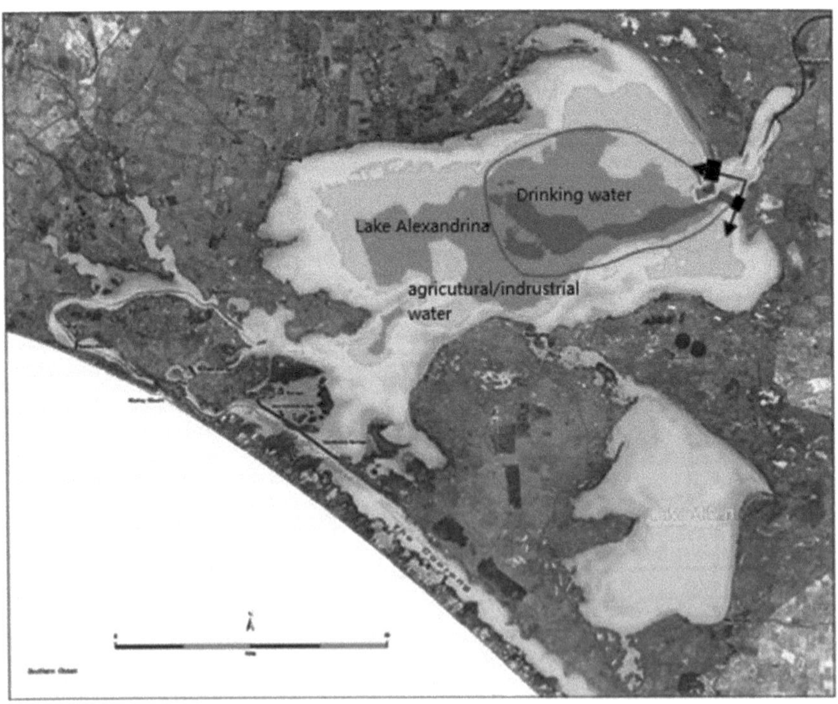

Figure 7.12 A coastal reservoir in Lake Alexandrina to develop freshwater from Murray–Darling River for Adelaide's water supply.

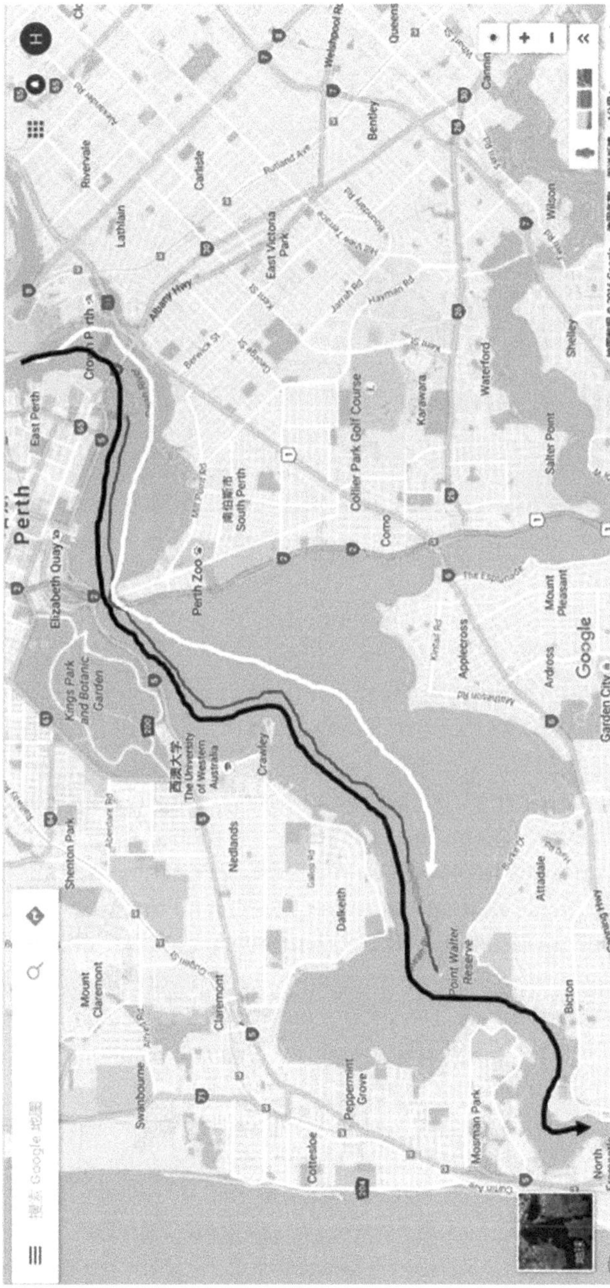

Figure 7.13 A coastal reservoir proposed at Swan River mouth for Perth's water supply.

Figure 7.14 Proposed coastal reservoir at the mouth of Adelaide River.

water supply can be secured. The Fullerton Cove (see Fig. 7.7) receives good-quality water in flood seasons, and its shallow cove can be used as a wetland to pretreat the river water that is then stored in the coastal reservoir in the sea. Fig. 7.8 shows that two barrages at the outlets of the Hawkesbury River mouth can create a reservoir of 50 km^2. These barrages can be also used for road transport such as freeway or high-speed train lines. If the average water depth is 10 m, then the storage is about 500 GL, whereas the annual flow rate is about 4600 GL/year. As the catchment is covered by forest, the water quality should be good enough, and the wastewater bypass channel may be unnecessary. Fig. 7.9 shows a coastal reservoir on the Shoalhaven River, whose flow is regulated by two hydraulic gates (0.15 and 0.5 km wide); these barrages can be also used for land transport. This coastal reservoir can be a backup water source for Sydney, Wollongong, Nowra and Canberra in extreme dry years in the future.

Melbourne's annual rainfall is about 650 mm/year. Its largest dam, the Thomson River dam, is capable of holding around 60% of Melbourne's water capacity. The total storage capacity of the 10 reservoirs that supply Melbourne is 1800 GL, and the water comes from 1600 km^2 of protected catchments in the Yarra Ranges. The annual flow from the catchment to these reservoirs is about 600 GL, but during 1997–2009, this dropped to

380 GL/year. On 1 July 2009, the storage was only 26% of full capacity. Water supply demand in 2014–15 was 400 GL. If coastal reservoirs are constructed at the river mouths of the Yarra River (catchment area = 4110 km^2), Bunyip River (4078 km^2), Maribyrnong River (1452 km^2) and Werribee River (1978 km^2), the total catchment is about 11,600 km^2, and the total run-off from these 4 catchments is estimates as 1970 GL/year. An example of a coastal reservoir for Melbourne is shown in Fig. 7.10 where the intake at the Yarra River is open only when the river flow has high-quality water; the lake in the city can be used as a wetland to pretreat the river water. The treated water is then stored in the coastal reservoir.

Adelaide's local inland reservoirs provided 32 GL/year water during the Millennium Drought (2000–09), and the catchments of these inland dams are small relative to the coastal reservoirs' catchment such as Torrens River and Onkaparinga River. Hence, if coastal reservoirs were constructed at these river mouths, the local reservoirs could provide more water to Adelaide, say 100–200 GL/year. In dry years, Adelaide's water supply depends heavily on water transfer from the Murray River, up to 90%. Lake Alexandrina was the first coastal reservoir in Australia as its barrages were constructed in the 1930s to develop its water resources. However, the design purpose was not successfully achieved as the Lake became a saline water body during the Millennium Drought.

Similar to Las Vegas, whose water supply comes from Hoover dam by developing water from the nearby Colorado River, Adelaide's desalination plants may be unnecessary if a dam like the Hoover dam existed to develop water from its nearby large river—the Murray–Darling River. Fig. 7.11 clearly shows that Adelaide is not short of water but is short of storage, as its nearby river's annual inflow during the Millennium Drought was 5700 GL/year, but the annual water demand was about 150 GL/year in the same period. Detailed research by Yang and Liu (2011) showed that if a second-generation coastal reservoir is applied inside Lake Alexandrina, this Australian 'Hoover dam' can supply sufficient high-quality water to the Adelaide and South Australia. Their results also showed that the coastal reservoir would improve the lake's ecosystem, as the research showed if there was a coastal reservoir as shown in Fig. 7.12 in the Millennium Drought, the lake's water level would have been higher and its hydraulic residence time would be shorter. This is understandable because the reservoir reduces the stagnant water volume and stores good-quality water in flood seasons. The research also showed that it is safe for Adelaide to divert 500 GL/year in dry years like 2007–09 from the coastal reservoir whose size is about 580 GL located inside the lake Alexandrina. In flood seasons, the intake gate is open to store high-quality water. After that, the intake gate is closed, and river flow by-passes the reservoir entering the lake and the poor-quality water is used to improve the ecosystem. This design does not use any shoreline around the lake; thus, it has little impact on human society.

Perth: Total population in West Australia in 2015 was 2.6 million, 82.7% lived in four coastal cities, i.e. Perth (2.0 million), Bunbury (0.075 million), Busselton (0.036 million) and Geraldton (0.040 million). In 2005, Perth had a population of around 1.5 million people, and the city consumed 240 GL/year of water. It is estimated that the population in Perth in 2026 and 2056 will be 2.27 million and 3.36 million, and the required water supply will be 350 and 520 GL/year, respectively. Run-off from Swan River (570 GL/year) and Murray River (720 GL/year, 42 km from Perth) can be developed using coastal reservoirs for Perth's water supply. An example is shown in Fig. 7.13, in which the bypass channel width is about 100 m on average, the estimated surface area of the reservoir is about 18.8 km^2, and the estimated storage capacity is about 100 GL. Alternatively, a barrage can be constructed near Point Walter Reserve and Chidley Reserve.

Darwin: The run-off from Adelaide River (1700 GL/year) in the Northern Territory can be developed for Darwin. The 180-km-long river is 50 km away from Darwin CBD. Its catchment is about 7640 km^2, and its tributaries include Margaret River, Manton River and Marrakai Creek. About 50% of the area is pastoral leasehold land, and 20% of it is Aboriginal land. The topography varies from low escarpment country in the upper reaches of Adelaide River to low lying flat floodplain areas near the coast. Fig. 7.14 shows a conceptual design of coastal reservoir for Darwin's water supply.

7.5 Interbasin water diversion for inland regions

Coastal reservoirs can supply sufficient water to coastal cities and also can provide sufficient water to inland Australia to meet the vast water demand for irrigation and mining industry. This is achievable as the existing large dams for water supply could be used for this purpose. For example, if the capital cities' water supply comes from the coastal reservoirs, then the water in large dams such as Warragamba dam, Wivenhoe dam and Thomson dam can be diverted to inland regions across the Great Dividing Range; thus, the

Murray–Darling River Basin's water could be increased. Water from some coastal reservoirs can be also diverted into the inland regions as shown in Fig. 7.15, where the proposed coastal reservoirs for water diversion are represented by blue ovals (dark grey ovals in printed version). The existing inland dams are indicated by red (grey in printed version) solid dots. The existing pipelines in SA and WA are represented by red lines (grey lines in printed version) without arrows. The proposed water diversion pipeline from Kimberley to Perth is represented by a blue line (dark grey line in printed version), and the proposed pipelines are represented by red lines (grey lines in printed version) with arrows. The cost and environmental impacts of this water diversion proposal could be horrendous beyond imagination.

To quantify the feasibility of these water diversion proposals, this study proposes that the energy required for pumping can be used to judge whether a water diversion route is feasible and economic by examining the pumping energy over long distance pipeline; the required energy can be empirically expressed as follows:

$$H(\mathrm{m}) = 0.625X(\mathrm{km}) + \Delta h(\mathrm{m}) \tag{7.1}$$

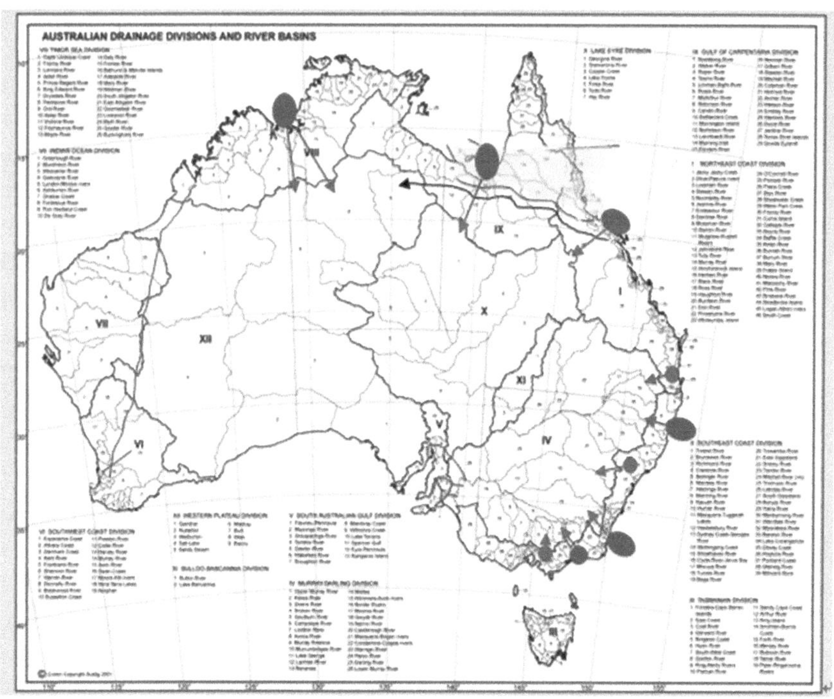

Figure 7.15 Proposed interbasin water diversion schemes from existing dams (*red arrow* [grey arrow in printed version] with *red cycles* [grey cycles in printed version]), and coastal reservoirs (*red arrow* [grey arrow in printed version] with *blue ovals* [dark grey ovals in printed version])

where H is the hydraulic head in m, and X is the total length of pipeline in km. Δh = elevation difference between the ending point and its source of route in m. Eq. (7.1) includes the energy loss by friction and the change of potential energy. In Table 7.3, the values of X (column 3) and elevations (columns 6 and 7) are obtained from Google Earth.

A feasible project can be seen from the existing water diversion pipelines in South Australia; the detailed information is shown in row 2 of Table 7.3. The Morgan-Whyalla pipeline is about 360 km long, Mount Lofty Ranges is about 460 m above sea level, and $\Delta h = 60-17 = 43$ m. Eq. (7.1) shows that $H = 268$ m, and the cost of water diversion is about AU\$0.24/m^3 (Marchi et al., 2014). The distance of water transfer from Wivenhoe dam across the Great Dividing Range is only X = 70 km lifting 384 m height; Eq. (7.1) gives $H = 428$ m. The calculated H using Eq. (7.1) is shown in the last column of Table 7.3.

Based on the H value, this study broadly divides the water division projects into four groups:

$H < 300$ m; class I;

301 m $\leq H <$ 600 m; class II;

601 m $\leq H <$ 1200 m; class III;

H $\geq$ 1200 m; infeasible.

Class I means that a project is simple and easy. A more difficult project may be defined as class II. A very difficult and costly project is defined as class III. If $H >$ 1200 m, the project is not feasible. From this criteria, the Morgan–Whyalla pipeline in South Australia is class I as its $H <$ 300 m. The project of Mannum–Adelaide pipeline is class II as its H is higher than 300 m but less than 600 m. The Perth–Kalgoorlie pipeline is class III (601 < H $\leq$ 1200 m). But the pipeline from Kimberley to Perth is not feasible as its H value is greater than 1200 m.

Based on the value of calculated H, feasibility of a proposed route of water diversion can be justified. Table 7.3 shows that water diversion from the coastal reservoir at the Carpentaria to Georgina River has $X = 250$ km and $\Delta h = 350$ m, and Eq. (7.1) gives $H = 506$ m, less than 600 m, so the project is feasible within class II. But Table 7.3 shows that it is infeasible to divert water from a coastal reservoir at the Bega River mouth where the distance X to the Great Dividing Range is only 50 km. Table 7.3 shows that the diversion routes are feasible but difficult (class III) from Wellington Lake and Warragamba dam to the Murray–Darling River system. Most of the other diversion routes shown in Fig. 7.15 are also feasible. However, more detailed investigation needs to be conducted for the feasibility and environmental impacts of these conceptual proposals.

7.6 Conclusions

This chapter investigates the available water resources and current countermeasures for its water crisis, and the following conclusions can be drawn:

(1) By analyzing the hydrological data, it is found that the annual run-off in Australia is about 400 km^3/year, but its water consumption is about 20 km^3/year, or 5%–6% of total run-off. Therefore, Australia is not short of water but short of water storage. This shortage is worsened with the time due to the rapid population growth. Based on the assumption that we are short of water, major projects such as desalination plants and wastewater reuse have emerged in Australia.

(2) This chapter reveals that the Australian water crisis is more severe than predicted. No new dams have been constructed since the 1990s for Australia's capital cities. Beside rapid growth of population, sedimentation keeps reducing Australian storage capacity. Consequently, in 2025, Australian storage per capita is only half of its peak capacity in 1985. By 2150, almost all existing dams in Australia would be out of service due to sedimentation and structural failures.

(3) If an additional 1%–2% of run-off is developed by coastal reservoirs, Australia may have sufficient water for its sustainable development. Australia was one of the pioneer countries in the world when it constructed a coastal reservoir for South Australia's water supply. This chapter analyzes why Lake Alexandrina with barrages cannot provide Adelaide sufficient clean water. An innovative design of a second-generation coastal reservoir has been proposed; applicability of this type coastal reservoir is discussed. The preliminary results indicate that by developing run-off lost to the sea, all the capital cities' water demands can be met even without inland dams or desalination plants.

(4) The research also shows that coastal reservoirs could provide sufficient water to Australian inland regions for its agricultural/industrial water needs. A new criterion is proposed to judge interbasin water diversion projects. This criterion reveals that the Kimberley-Perth and Bega River–Murray River pipelines are not feasible, and both had received hot debate during the Millennium Drought. However, once the coastal cities' water supply comes from coastal reservoirs, water in the existing inland reservoirs can be diverted to Australian inland regions. As Australia is a flat country, water from some inland/coastal reservoirs can also be economically diverted to the central region.

Although the preliminary research shows that coastal reservoirs are an environmentally sustainable, cost-effective and technically feasible technology, more detailed research is needed to justify the conceptual designs shown in this chapter.

References

Bae, Y.H., Kim, K.O., Choi, B.H., 2010. Lake Sihwa tidal power plant project. Ocean Engineering 37 (5–6), 454–463.

Batuca, D.G., Jordaan, J.M., 2000. Silting and Desilting of Reservoirs. Balkema Publishers, Rotterdam, The Netherlands.

Bureau of Meteorology, 2015. Water in Australia 2013-14. Bureau of Meteorology, Melbourne, Australia. Australian government report.

Chu, J., Yan, S.W., Li, W., 2012. Innovative methods for dike construction—an overview. Geotextiles and Geomembranes 30 (2), 35–42.

Finlayson, B.L., McMahon, T.A., 1988. Australia v. the world: a comparative analysis of streamflow characteristics. In: Warner, R.F. (Ed.), Fluvial Geomorphology of Australia. Academic Press, Sydney, pp. 17–40.

Lahnsteiner, J., Lempert, G., 2007. Water management in Windhoek, Namibia. Water Science and Technology 55 (1–2), 441–448.

Lammens, E., van Luijn, F., Wessels, Y., Bouwhuis, H., Noordhuis, R., Portielje, R., van der Molen, D., 2007. Towards ecological goals for the heavily modified lakes in the IJsselmeer area, The Netherlands. Hydrobiologia 599, 239–247.

Larsen, J., Gibbes, B., Quiggin, J., 2014. Dam Hard: Water Storage is a Historic Headache for Australia. The Conversation. October 27th. http://theconversation.com/dam-hard-water-storage-is-a-historic-headache-for-australia-33397.

Marchi, A., Dandy, G., Maier, H., 2014. Financial Costs, Energy Consumption and Greenhouse Gas Emissions for Major Supply Water Sources and Demand Management Options for Metropolitan Adelaide. Goyder Institute for Water Research. Technical Report Series No. 14/12.

Puckridge, J.T., Sheldon, F., Walker, K.F., Boulton, A.J., 1998. Flow variability and the ecology of large rivers. Marine and Freshwater Research 49, 55–72.

Seqwater, 2015. Seqwater Annual Report 2014–2015. Queensland Bulk Water Supply Authority (Brisbane, Australia).

Stamatov, V., Stamatov, A., 2010. Long-term impact of water desalination plants on the energy and carbon dioxide balance of Victoria, Australia: a case study from Wonthaggi. Water and Environment Journal 24, 208–214.

Turner, A., Hausler, G., Carrard, N., Kazaglis, A., White, S., Hughes, A., Johnson, T., 2007. Review of water supply-demand options for south east Queensland. Water Engineering Australia 2 (8), 30–37.

U.S. Bureau of Reclamation (USBR), 2006. Erosion and Sedimentation Manual. U.S. Department of the Interior, Bureau of Reclamation, Technical Service Center, Sedimentation and River Hydraulics Group, Denver, Colorado.

White, R., 2001. Evacuation of Sediments from Reservoirs. Thomas Telford, London.

World Bank, 2019. A Water-Secure World for All. World Bank, Washington, DC.

Yan, S.W., Chu, J., 2010. Construction of an offshore dike using slurry filled geotextilemats. Geotextiles and Geomembranes 28 (5), 422–433.

Yang, S.Q., 2003. Potential water resources in Singapore. Journal of Water Supply: Research and Technology 52 (6), 425–434.

Yang, S.Q., 2004. Global and China's Water Crisis and its Solutions in the 21st Century. Tianjin Univ. Press, Tianjin. ISBN 7-5618-2069-0 (in Chinese).
Yang, S.Q., 2015. A preliminary feasibility study for a backup water supply coastal reservoir in Southeast Queensland, Australia. Journal of Water Supply: Research and Technology 64 (4), 470–485.
Yang, S.Q., 2016. Why may coastal reservoir dominate future water supply?. In: Proceedings of ANCOLD/NZSOLD 2016 Conference, Adelaide, October 18–19, 2016.
Yang, S.Q., Liu, P.W., 2010. Strategy of water pollution prevention in Taihu Lake and its effects analysis. Journal of Great Lakes Research 36 (1), 150–158.
Yang, S.Q., Lin, P.Z., 2011. Coastal reservoir by soft-dam and its possible applications. Recent Patents on Engineering 5, 45–56.
Yang, S.Q., Zhang, Z., Ng, B.C., Ong, B.C., 2001. Coastal Reservoirs. Singapore Patent No, pp. 200504653–200504657.

CHAPTER 8

Coastal reservoirs and their potential for urban regeneration and renewable energy supply

Roger A. Falconer, Bin Guo, Reza Ahmadian
Hydro-environmental Research Centre, Cardiff University, Cardiff, United Kingdom

8.1 Introduction

The world faces considerable water management challenges now and increasingly in the future, primarily brought about by: the predicted and anticipated impacts of climate change, the increasing need to provide more water, food and energy for a growing global population, and increasing globalization, leading to a more inequitable global wealth distribution and a corresponding increase in demand for water, associated with the need for more commodities (such as cars, clothes etc.). In addition, in most countries worldwide there is an increasing population shift away from rural communities to larger urban cities, where there are better employment prospects, and with these cities often being located near the coast. However, coastal cities are also frequently vulnerable to limited access to adequate water resources, poor water sanitation, coastal and estuarine flooding, sea level rise, increased energy demands, etc. It is therefore timely that we look to the future and address some of these challenges in a sustainable manner and working with nature to develop coastal reservoirs to contribute to the solution of some of these challenges.

Coastal reservoirs have recently been primarily considered as a means of providing a cost-effective solution for storing fresh water toward the end of a river basin, enabling the fresh water to be treated and distributed around the city at a lower cost than such alternatives as building high head reservoirs, upstream along the river basin, and then having to pump the treated water to the city for distribution. The water level can also be controlled in coastal reservoirs, thereby also offering opportunities to reduce flood risk from the sea locally and abstract more water during river floods, thereby lowering upstream water elevations, and hence reducing flood risk to coastal cities. Furthermore, by taking water from the estuary at the end

Sustainable Water Resource Development Using Coastal Reservoirs
ISBN 978-0-12-818002-0
https://doi.org/10.1016/B978-0-12-818002-0.00008-3

of the river basin, rather than from an impounded reservoir some distance up the river basin, then the river flow is not reduced disproportionately during drought conditions and the river ecosystem and low environmental flows are less affected by water abstraction etc.

In this chapter details are given of two related studies involving the design and operation of coastal reservoirs for very different primary purposes, other than the more traditional use of coastal reservoirs for water supply. These two reservoir studies include: (1) the Cardiff Bay project, where a tidally flushed estuarine bay in South Wales was closed off with a barrage, creating an impounded freshwater reservoir for urban regeneration; and (2) the West Somerset Lagoon proposed project, where a tidal lagoon is being considered and designed to provide tidal renewable energy in a macro-tidal estuarine basin, namely the Bristol Channel, in the southwest of the United Kingdom.

For the Cardiff Bay reservoir project, the port of Cardiff grew rapidly in the second half of the 19th century, reaching maximum growth at the turn of the century (i.e., 1850–1900) and with the port reaching maturity around 1920. After the loss of trade to deeper ports, around 1920, the port went into rapid decline, and by 1980 the city of Cardiff, capital city of Wales, was approximately 1–2 km inland of the port and with wasteland up to the bay waterfront. In the 1980s a 1.1 km tidal exclusion barrage was proposed across the mouth of the bay, creating 13 km of waterfront for Cardiff and impounding two rivers, the Taff and Ely. The project was approved and funded by the UK government through the Cardiff Bay Barrage Act in 1993, and the barrage was completed and the project opened in 1999. The primary aim of this coastal reservoir was to regenerate the area and re-connect Cardiff to its waterfront along the macro-tidal Severn Estuary. Further details of the background to this project are given in Crompton (2002).

The freshwater reservoir is now very different from the previous tidally flushed bay and some of the key hydro-environmental challenges can be summarized as follows: (1) the bay now has long retention times, whereas previously it had experienced the second highest tidal range worldwide; (2) combined sewer overflows discharge directly into the rivers and then into the bay during flood events; (3) nutrient and animal diffuse source inputs discharge into the rivers (particularly the Ely) and then into the bay during storm events; (4) contaminated sediments from past industries and port activities have the potential to release toxic substances into the reservoir; and (5) low levels of dissolved oxygen in the lower layers of the reservoir during summer months could be problematic due to stratification. In addressing these challenges and concerns, extensive field, laboratory and numerical modelling studies were undertaken, with the latter two studies being undertaken within the Hydro-environmental Research Centre (HRC) at Cardiff University; these studies are summarized herein.

Coastal reservoirs are also appropriate for tidal energy generation where ideal conditions exist, such as a large tidal range and suitable geographical conditions. Such reservoirs, or lagoons, offer the potential possibility to be designed with multi-function features in addition to their main function of tidal range energy generation. Tidal range schemes (TRSs), such as barrages or lagoons, are built by blocking off the entire estuary or part of the estuary and/or coast. TRSs have several advantages compared to other renewable energy generation schemes, including high predictability and long-life cycle (120 years). Moreover, they could be multi-functional projects and be used as freshwater storage, flood alleviation and regeneration and tourism.

TRSs are operated in such a manner as to create an artificial head difference across the impoundment wall and then to make use of this head difference to generate electricity. The artificial head difference is created through holding water inside the basin at high and low tides and is therefore influenced by the tidal range. The amount of electricity generated is directly proportional to the square of this artificial head difference and the tidal range at the location of the scheme, with more information being given in Baker (1991) and Falconer et al. (2009). Therefore, sites with a high tidal range and a large impounded area are potentially particularly attractive for TRSs. The Severn Estuary and Bristol Channel has the second highest tidal range in the world and is one of these locations, with several schemes including different variations of the Severn Tidal Power Group (STPG) Severn Barrage, Swansea Bay Lagoon, Fleming Lagoon and, more recently, West Somerset Lagoon being proposed in this large estuarine basin, as discussed by Falconer et al. (2009), Ahmadian et al. (2010a,b), Čož et al. (2019) and Xia et al. (2010). Due to the nature of TRSs, the impact of such coastal reservoirs or lagoons on the hydrodynamics, water quality, sediment transport and morphodynamics, and ecology, within and around the schemes, must be thoroughly investigated to avoid potential adverse hydro-environmental impacts. The second part of the chapter therefore highlights the hydro-environmental modelling studies of the West Somerset Lagoon and impact of the scheme on retention times inside the lagoon.

8.2 Numerical model details

The main numerical model used for predicting the hydrodynamic and water quality indicator levels in the studies reported herein was based on the Hydro-environmental Research Centre's 3D numerical model, TRIVAST. The model first solves the three-dimensional Navier–Stokes equations written for mass conservation and momentum as:

$$\frac{\partial u}{\partial x}+\frac{\partial v}{\partial y}+\frac{\partial w}{\partial z}=0 \tag{8.1}$$

$$\frac{\partial u}{\partial t}+\frac{\partial u^2}{\partial x}+\frac{\partial uv}{\partial y}+\frac{\partial uw}{\partial z}=+fv-\frac{1}{\rho}\frac{\partial P}{\partial x}-\left[\frac{\partial \overline{u'u'}}{\partial x}+\frac{\partial \overline{u'v'}}{\partial y}+\frac{\partial \overline{u'w'}}{\partial z}\right] \tag{8.2}$$

$$\frac{\partial v}{\partial t}+\frac{\partial uv}{\partial x}+\frac{\partial v^2}{\partial y}+\frac{\partial vw}{\partial z}=-fu-\frac{1}{\rho}\frac{\partial P}{\partial y}-\left[\frac{\partial \overline{u'v'}}{\partial x}+\frac{\partial \overline{v'v'}}{\partial y}+\frac{\partial \overline{v'w'}}{\partial z}\right] \tag{8.3}$$

$$\frac{\partial P}{\partial z}+\rho g=0 \tag{8.4}$$

where u, v, w = velocity components in the x, y, z co-ordinate directions, respectively, t = time, f = Coriolis parameter, ρ = fluid density, P = the fluid pressure, $\overline{u'u'}$, $\overline{u'v'}$, $\overline{u'w'}$ = Reynolds stresses in x direction on x, y, z planes, respectively. For the Reynolds stresses these terms were solved using a two-layer mixing-length model. For further details of the mixing-layer model and the three-dimensional formulation, see Lin and Falconer (1997) and Falconer and Lin (1997).

The governing Eqs (8.1)–(8.4) were solved using a finite difference technique on a square mesh, covering the region of interest in the horizontal plane and a layer integrated finite volume scheme in the vertical plane. The number of grid squares varied according to the depth across the domain, with the option to set the grid size as regular or irregular in the vertical. The depth integrated equations were used to determine the water surface elevation field using the layer integrated velocities. The layer integrated equations were then used to determine the velocity field using the corresponding water elevation field. The alternating direction implicit method was modified to solve the depth integrated equations. A description of the bathymetry and bed roughness was given for each grid square and water elevations or currents included along the open boundaries.

In numerically modelling the flux of water quality indicator organisms, the conservation of solute mass equation was solved, with the three-dimensional form of the equation being given as:

$$\frac{\partial \phi}{\partial t} + \frac{\partial \phi u}{\partial x} + \frac{\partial \phi v}{\partial y} + \frac{\partial \phi w}{\partial z} + \frac{\partial}{\partial x}\overline{u'\phi'} + \frac{\partial}{\partial y}\overline{v'\phi'} + \frac{\partial}{\partial z}\overline{w'\phi'} = \phi_S + \phi_d \quad (8.5)$$

where φ = time averaged faecal solute concentration, φ_S = source or sink solute input (e.g., an outfall), and φ_d = faecal coliform decay term.

The cross-produced terms $\overline{u'\phi'}$ etc. represent the mass flux of solute due to the turbulent fluctuations and, by analogy with Fick's law of diffusion, can be assumed to be proportional to the mean concentration gradient and in the positive direction of decreasing concentration. Hence, the terms can be written as:

$$\overline{u'\phi'} = -D_{tx}\frac{\partial \phi}{\partial x} \quad (8.6)$$

where D_{tx}, etc. = turbulent diffusion coefficient in x direction etc. These coefficients are often associated with the eddy viscosity ε_t by a Schmidt number, with its value found to vary between 0.5 and 1.0.

For the variation of concentration distributions of faecal coliform, the solute transport equation was integrated over a layer of the water column depth with the inclusion of turbulent diffusion, source and sink inputs and decay rate constants. In numerical model simulations relating to the transport of faecal coliform indicator organisms, the solute mass conservation equation was often applied to, and solved for, faecal coliform distribution fields where high concentration gradients exist. The numerical representation of the advective terms in the solute mass conservation equation are critical in terms of the degree of numerical diffusion introduced into the scheme and the occurrence of grid scale oscillations, or undershoot and overshoot, arising in the vicinity of large concentration gradients. Cayhono (1993) undertook an extensive study in comparing 36 of the most popular finite difference schemes for modelling high concentration gradients, with comparisons showing that the ULTIMATE scheme, after Leonard (1991), was particularly attractive since it was more general than the other schemes considered and was relatively easy to apply. The ULTIMATE and the third-order QUICKEST schemes were combined to represent the advective terms in the solute mass conservation equation. This algorithm reduces considerably the introduction of numerical diffusion and eliminates undershoot or overshoot in regions of high concentration gradients. This factor is particularly important for point sources (such as the riverine inputs into Cardiff Bay) and the episodic discharges from combined sewer overflows.

For the faecal coliform decay rate, this term is generally written in most numerical models as a simple first-order decay rate. However, extensive reviews of the literature show that this kinetic process is at least a function of: sunlight intensity, irradiance, depth, pH, temperature and suspended particulate matter. In particular, the work of Gameson and Saxon (1967) showed that the faecal coliform decay rate was highly dependent upon light intensity. The identification of sunlight as a major bactericidal agent has many ramifications with the expected survival rate of sewage-borne bacteria in the natural environment. Bellair et al. (1977) undertook a series of experiments that commenced at 6:30 a.m., when it was found that the die-off was low. However, it was found to increase rapidly, reaching a maximum around noon. Recorded values ranged from 19 to 40 h, with the inactivation rate varying greatly, particularly over a diurnal cycle, and with the rate of die-off at any time of the year being approximately proportional to the intensity of irradiance received by the sample. The effect of sunlight on the mortality of faecal coliform die-off was approximated by being related to the irradiance through a power law function as given by:

$$k_s = \alpha I^{\beta} \tag{8.7}$$

where k_s = die-off rate due to sunlight intensity (day^{-1}), I = irradiance (W/m^2), α = constant of proportionality and β = slope of the log plot of die-off against irradiance. The degree of sunlight penetration into the water column has a significant effect on the bacterial die-off beneath the water column. The turbidity of the water interferes with the light penetration through the water column and affects the bactericidal effectiveness of sunlight. Therefore, in more turbid waters the bacteria survival time is increased, mainly because of the decreased effect of UV light, which is partially adsorbed by the suspended matter. The penetration of incoming solar irradiation is usually described by an extinction coefficient. This coefficient is proportional to the water depth and may be calculated from solar radiation measurements taken at a range of water depths and represented by the Lambert law, given as:

$$I_z = I_0 e^{-k_e t} \tag{8.8}$$

where I_0 = irradiance at the surface (W/m^2), I_z = irradiance at depth z (W/m^2), z = depth (m) and k_e = vertical light extinction coefficient (m^{-1}).

To enable the inclusion of a dynamic method for determining the survival rates of bacteria in the numerical model, two field measurement exercises were undertaken to determine the key environmental parameters affecting the faecal indicator organism levels in the bay and the rivers Taff and Ely. These surveys were conducted in March (characterizing cold, overcast conditions) and July (characterizing hot, sunny conditions). The surveys involved the deployment of two survey boats within the bay, as well as land-based survey teams to sample the rivers and other inputs. The surveys were conducted over the period dawn to dusk, to provide information on the diurnal variability of relevant parameters. The main parameters measured, at various depths, included bacterial concentrations, solar irradiance in air and water, water temperature, turbidity, pH, salinity, conductivity, suspended particulate matter and water depth. The hydro-environmental model was set up with field values of the vertical light extinction coefficient for the March and July surveys and the die-off rate directly related to the solar intensity, implying that the die-off rate in darkness would be zero. Gameson and Gould (1975) reported that the effects of sunlight on coliform die-off were additive and independent of temperature, hence the die-off rate was expressed as the sum of the die-off rate for darkness, k_d, and the die-off rate due to sunlight, k_s. If the total faecal coliform mortality rate, k, can be defined by a simple relationship, considering darkness and sunlight mortality, then this relationship gives:

$$k = k_d + k_s \tag{8.9}$$

where typical values for k_d and k_s were 100 and 10 h^{-1}, respectively. Extensive further research was undertaken by Kashefipour et al. (2002) and Lin et al. (2003) where extensive field data of faecal coliform levels were used, together with artificial neural networks, to develop new formulations for the faecal coliform decay rate k.

8.3 Cardiff Bay study

8.3.1 General considerations

Cardiff Bay Barrage, and the impounded freshwater reservoir, provides an excellent example of a coastal reservoir designed and built in the United Kingdom for urban regeneration. The project, opened in 1999, focused on reconnecting Cardiff with its waterfront. The original port of Cardiff was

Figure 8.1 Photograph of Cardiff Docks taken at its peak around 1900 (BBC). *(From https://commons.wikimedia.org/wiki/File:Coal_ships_tied_up_at_Cardiff_Docks.jpg.)*

located adjacent and around the waterfront of Cardiff Bay, with the city being located about 1–2 km north and inland from the waterfront. During its heyday (c.1920) the port of Cardiff was one of the largest in the world, as shown in Fig. 8.1 (http://www.cardiffbay.co.uk/history/). From the 1920s onwards the port went into decline; there was a major loss of trade to deeper ports around the United Kingdom, as the ships became larger and with the increasing challenges associated with the Bristol Channel and Severn Estuary having the second largest tidal range in the world. In 1999 a barrage was completed across the mouth of the tidal bay and the Cardiff Bay reservoir was officially opened as a freshwater lake and an urban regeneration project, see Fig. 8.2.

The main objective of the Cardiff Bay Barrage act of Parliament was to regenerate the area surrounding Cardiff Bay and reconnect the city of Cardiff with its waterfront. This included: (1) a tidal exclusion barrage of length 1.1 km; (2) a new 13 km of waterfront for the city of Cardiff; and (3) impoundment of two rivers, namely the Taff and Ely, to create a 200 ha freshwater reservoir. The key objectives of the regeneration included: (1) creating 30,000 new jobs, 1M m^2 of industrial development and 6000 new homes; (2) a strategy underpinning a tourism potential of attracting two million visitors annually to Cardiff; and (3) maintaining high-quality water in the reservoir, with a legal requirement of a minimum dissolved oxygen

Figure 8.2 Cardiff Bay: a coastal reservoir and urban regeneration project.

level of 5 mg/L, as required by the Cardiff Bay Barrage Act 1993. The 1.1 km-long barrage included an earth embankment, five sluices, three locks, an outer harbour and a fish pass, with details of the scheme being given in Hunter and Gander (2002).

With the change in the characteristics of Cardiff Bay from a tidally flushed basin to an impounded freshwater reservoir, there have been several key challenges to consider, and particularly regarding the hydrodynamic, water quality, sediment transport and morphological processes. These challenges were introduced earlier in the paper, but include the following key hydro-environmental impacts: (1) changes in the hydrodynamics and much longer flushing times in the bay; (2) combined sewer overflows discharging into the rivers and reservoir during flood events; (3) nutrient and animal diffuse source inputs discharging into the rivers and reservoir; (4) contaminated sediments from past industries increased sedimentation in the reservoir; and (5) potentially low dissolved oxygen levels and stratification in the reservoir lower layers during summer months. In studying many of these hydro-environmental challenges, several research studies were undertaken by the authors, in the HRC at Cardiff University. These studies mainly focused on numerical modelling studies, but also included small-scale physical model studies in the Hyder Hydraulics Laboratory at Cardiff University and field data led by the Centre for Research in Environment and Health at Aberystwyth University.

8.3.2 Modelling details

As a result of the concerns outlined above, the authors in the HRC at Cardiff University undertook several research studies to refine and apply an integrated modelling tool for water quality management of the bay. The project strategy consisted of integrating a CSO model (namely SWMM), with a 1-D river model (namely FASTER) and a 3D hydro-environmental numerical model (namely TRIVAST). The latter two models were developed by the HRC, whereas the model SWMM is a storm water rainfall-runoff model developed by the U.S. Environmental Protection Agency. An example of such a research study involved investigating the impact of a barrage and the creation of the freshwater coastal reservoir on the faecal indicator organism (FIO) levels in the bay, before and after construction of the barrage. The numerical modelling studies were complemented with an extensive field monitoring programme to establish more precise values for the kinetic decay rates in the bay, for various meteorological conditions (such as different sunlight intensities), thereby providing improved predictions of FIO levels across the bay.

The numerical model was set up for the bay and riverine inputs, using a regular grid of 50 m cell size, and with the 3D model having 10 vertical layers, each of height 1.25 m. The downstream boundary was governed by the barrage and its operating conditions. The water level in the reservoir was controlled by the barrage operation to maintain a near-constant level of between 4.0 and 4.5 m above ordnance datum (OD), with the flow through the barrage being governed by the sluices when the tidal elevation in the Severn Estuary externally was lower than the impounded water level and with the flow governed by pumps when the estuary elevation was above the impounded level. For the upstream boundary conditions the flows in the rivers Taff and Ely were included, either as mean values or time-varying hydrographs. The mean river flow in the Taff was found to be 20 m^3/s and with a peak flow of 800 m^3/s; likewise, the flow in the Ely was typically 20% of that in the Taff. The bottom roughness parameter was treated as a constant across the domain and with a calibrated value in the form of a roughness height k_s, equating to an equivalent Manning roughness coefficient of approximately $n = 0.03$.

Simulations were first undertaken to validate the numerical model by predicting the movement of an arbitrary spillage of a conservative tracer with zero decay into the rivers Taff and Ely and by comparing the numerical model predictions with small-scale laboratory observations.

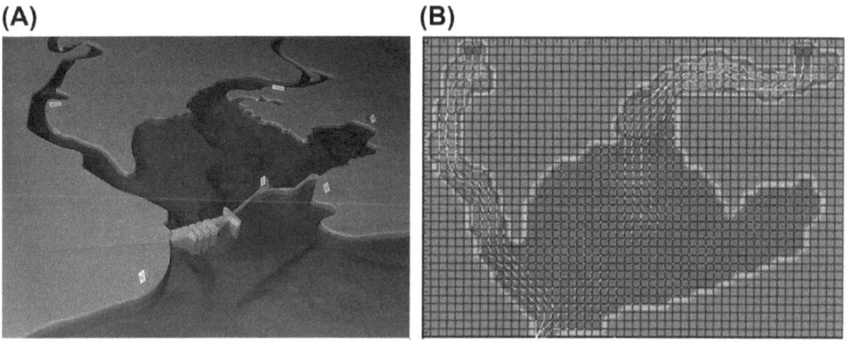

Figure 8.3 Comparison of observed conservative tracer plume for: (A) scaled laboratory model, and (B) numerical model predictions.

The small-scale physical model of Cardiff Bay was sited in a tidal basin, in the Hydraulics Laboratory at Cardiff University. Scaled tidal elevations were governed through an oscillating overflow weir, being driven by computer and generating tides of varying or constant amplitude and period. The physical model also included a model barrage at the mouth of the bay to replicate the impounded waters, with the presence of the barrage preventing the tidal influence from affecting the waters inside the bay. The physical model had a distorted scale, with the horizontal and vertical scale ratios being 1:1500 and 1:75, respectively, with a much larger scale, 1:250, physical model study being undertaken at HR Wallingford for this project and reported by Burt (2002). A volume of tracer was injected into the rivers and the trajectory of the plumes observed accordingly. Fig. 8.3A shows the measured and observed plume trajectories in the scaled physical model, with comparisons with the numerical model results shown in Fig. 8.3B. As can be seen by comparing Fig. 8.3A and B, the numerical model and physical model plume trajectories are in close agreement. Further details of the modelling are given in Harris et al. (2002).

8.3.3 Model results

The model was then set up to study the impact of episodic inputs into the rivers and for a range of wind and riverine flow conditions. Some of these predictions were compared with the field data, with the corresponding comparisons for different boundary conditions enabling a comprehensive set of conclusions to be drawn to indicate water management strategies for coastal reservoirs for the future. As an example of these research studies, simulations were undertaken to investigate the impact of using dynamically measured varying decay rates for FIO levels in the receiving waters in the rivers and reservoir, as a result of point- and diffuse-source pollutant inputs. It is known that bacterial decay is a function of several variables, including temperature, sunlight intensity, solar irradiance, depth, pH and suspended particulate matter. In particular, bacterial decay was first investigated for sunlight intensity, and field measurements were undertaken in the bay by the Centre for Environment and Health, led by Professor David Kay and reported in Stapleton et al. (2002), and Kay et al. (2005), in March (including cold and overcast conditions) and July (including hot and sunny conditions). The surveys involved the deployment of two survey boats within the bay, as well as land-based survey teams, with data being monitored at the sites shown in Fig. 8.4. The surveys were conducted from dawn to dusk, providing information on the diurnal variability of several key parameters. The main parameters measured at various depths included: faecal coliform and faecal streptococci levels, solar irradiance, salinity, conductivity, temperature, pH and depth.

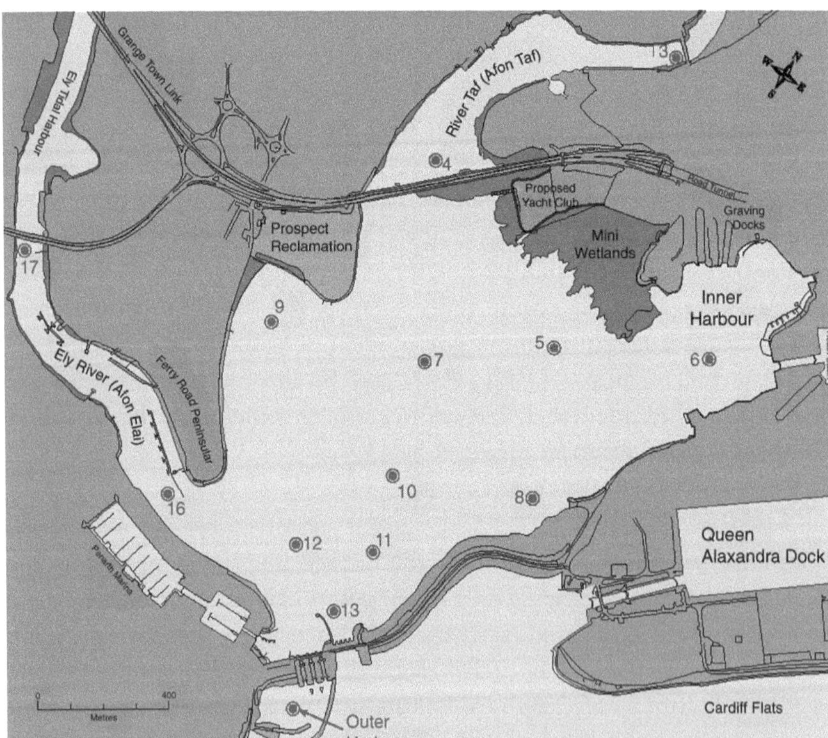

Figure 8.4 Field monitoring sites in the rivers Taff and Ely and Cardiff Bay reservoir.

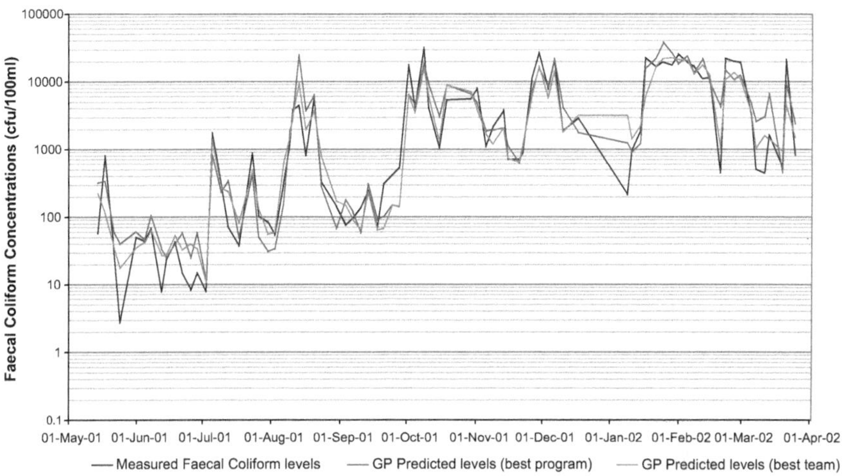

Figure 8.5 Comparison of measured and predicted faecal coliform levels at Site 5 using genetic programming.

The two field data studies included comparisons being made of the impounded FIO levels for a range of the variables outlined above, with the results showing that the FIO levels were highly dependent upon several of these key variables. From these data a genetic algorithm was used to refine Eq. (8.9) to give improved predictions of the decay rate as a function of sunlight intensity, see Harris (2003). A typical example of the analysis of the data using a genetic algorithm is shown in Fig. 8.5, with the corresponding formulation being given as follows:

$$k = 2.63 + 0.24\, I^{0.63} \tag{8.10}$$

where k = dynamic decay rate (day^{-1}). The three-dimensional hydro-environmental model was then run for hypothetical spillages into the rivers and initially for a constant decay rate (T_{90}), both for day- and nighttime, of 60 h. Subsequently, the model was re-run for daytime and nighttime decay rates obtained using the formulation given in Eq. (8.10) above, with the corresponding T_{90} (i.e., time for 90% decay, specified in hours) values varying from approximately 10—100 h for day- and nighttime, respectively. The difference in the predictions was significant and indicated that, for this freshwater basin, nighttime spillages during the autumn and winter months led to reduced faecal coliform levels in the rivers and bay, in comparison with corresponding spillages occurring during the notional 12 h daytime period, see Fig. 8.6, with further details given in Harris et al. (2002).

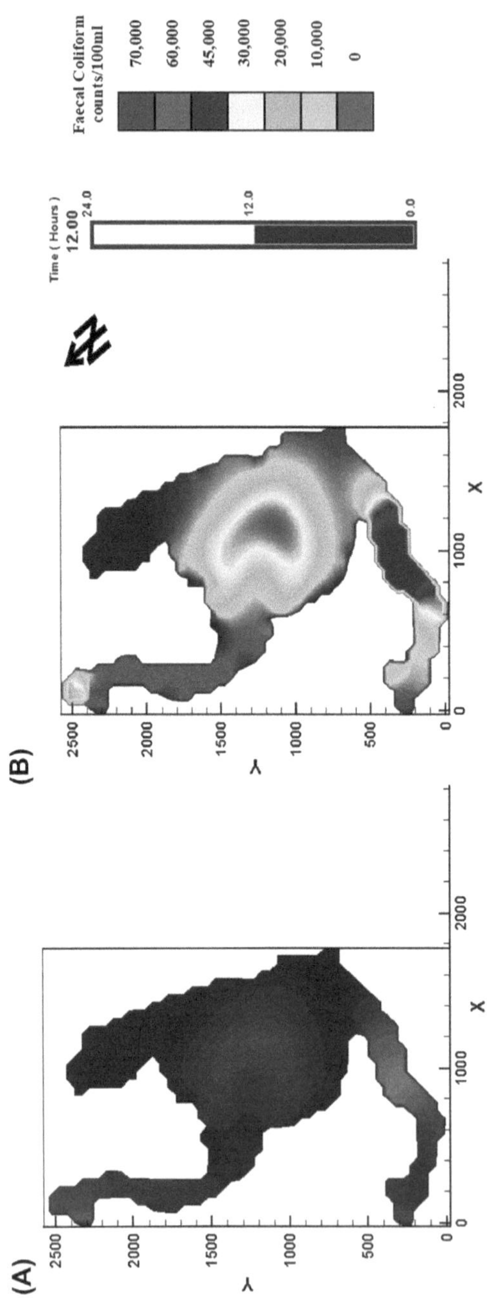

Figure 8.6 Faecal coliform distributions for arbitrary releases into rivers Taff and Ely at: (A) 4 a.m. and (B) 4 p.m.

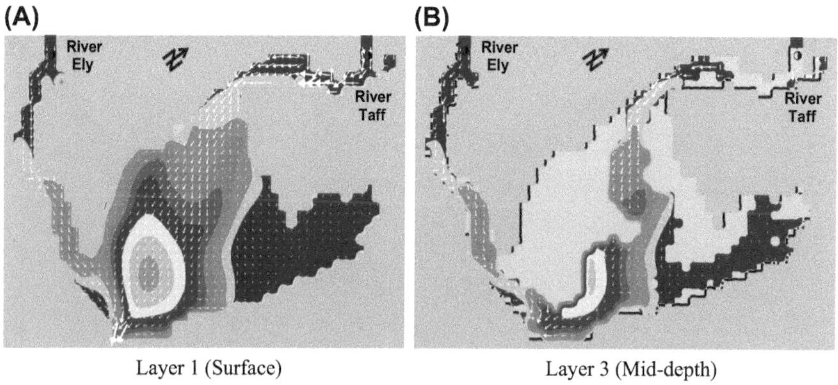

Figure 8.7 Three-dimensional model predictions of the surface and third-layer hydrodynamics and FIO levels for arbitrary input through the rivers.

Further field measurements and model predictions were undertaken, with the die-off related to sunlight intensity, temperature and irradiance.

Finally, a series of simulations were undertaken using the 3D hydro-environmental model TRIVAST (ThRee-dimensional Integrated Velocities And Solute Transport) to predict the FIO levels within the layers arising from arbitrary inputs coming into the reservoir via diffuse- and point-source inputs into the rivers. The model showed limited indications of stratification and faecal coliform levels were traced throughout much of the water column depth across the reservoir, as indicated in Fig. 8.7. The dissolved oxygen concentrations within Cardiff Bay are maintained at or above 5 mg/L through an extensive aeration system across the bay. Compressed air is pumped, from five sites around the bay, through a series of steel-reinforced rubber pipelines, laid on the beds of the bay and the rivers Taff and Ely. These are connected to over 300 diffusers. At times this aeration system is insufficient to maintain the dissolved oxygen levels in the lower layers of the reservoir and the Harbour Authority then uses a mobile oxygenation barge with liquid oxygen in a tank. The barge can dissolve up to 5 tonnes of oxygen into the reservoir in 24 h, with this additional facility maintaining the dissolved oxygen level about 5 mg/L in the critical summer months and avoiding the onset of stratification.

8.4 West Somerset Lagoon study

8.4.1 General background

The proposed West Somerset Lagoon (WSL) spans the southern coast of the Bristol Channel as shown in Fig. 8.8A. In the preliminary design, the

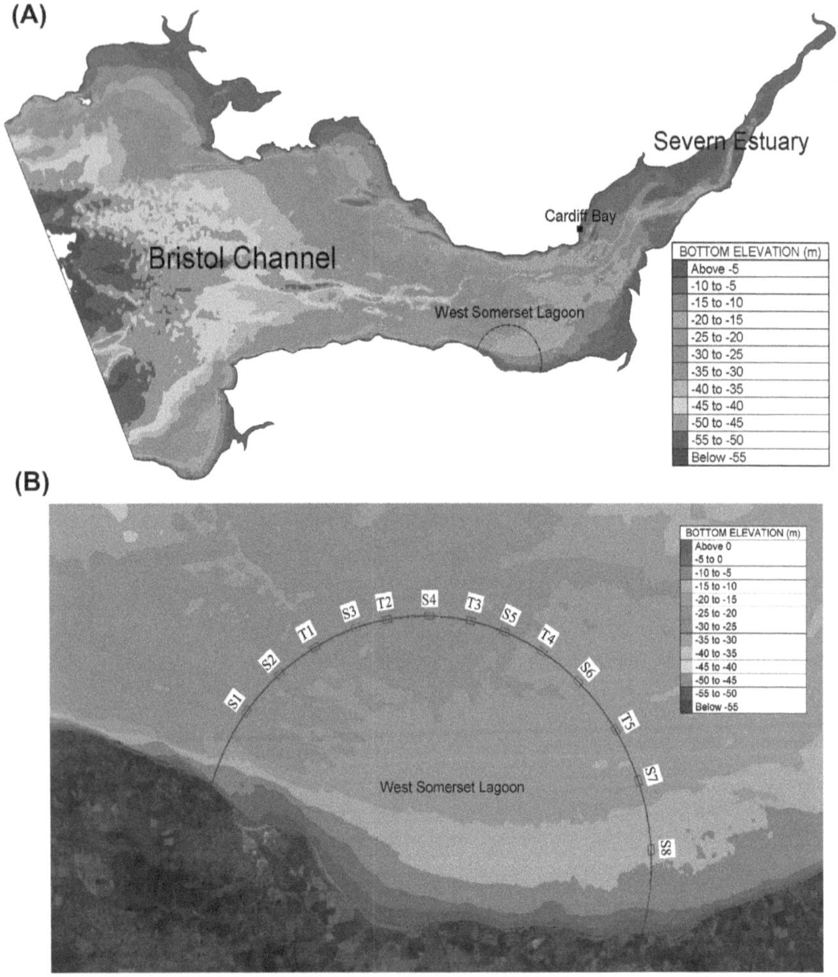

Figure 8.8 Illustration of: (A) the Bristol Channel; and (B) the layout of West Somerset Lagoon, including distribution of the turbine and sluice gate blocks.

WSL proposal includes a breakwater, with a length of 21 km, which impounds a coastal area of 72 km^2. As shown in Fig. 8.8B, T1–T5 represent five turbine blocks and S1–S8 illustrate the locations of the proposed sluice gates. The 125 bulb turbines, each of diameter of 7.2 m, are split into five blocks. The blocks are distributed uniformly, taking account of the local bathymetry and geological conditions in the design. The capacity of each turbine is 20 MW, which gives the scheme a total power capacity of 2.5 GW. To enhance the operation, eight blocks of sluice gates

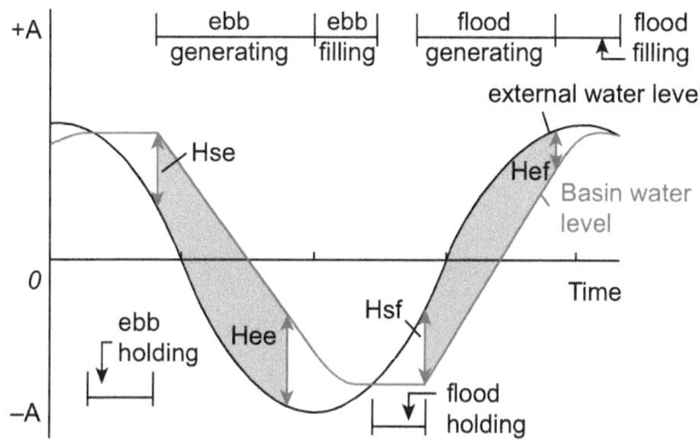

Figure 8.9 Schematic representation of two-way operational schemes.

of two different sizes are planned. The details of the scheme are shown in Fig. 8.8. The sluice area of each block in S1 to S5 is 2860 m^2, with each block in S6 to S8 having a sluice area of 1900 m^2. The total sluice area of the proposed scheme is 20,000 m^2. The proposed operation for the WSL scheme is two-way generation, as shown in Fig. 8.9, which has been shown to be the more favourable operation, based on the research results of Xia et al. (2010), Ahmadian et al. (2014) and Xue et al. (2019).

Starting at the high tide, the turbines and sluice gates remain closed to keep water levels constant inside the lagoon, while the water levels outside the lagoon are falling. This phase, which is referred to as the holding phase, continues until the head difference across the lagoon reaches the prescribed head difference, i.e., the starting head, where the ebb-generation phase starts. The turbines will operate and generate electricity in this phase, while the sluice gates are closed. The generation phase is followed with the sluicing phase, when the head difference across the lagoon becomes smaller than the end head difference. In the sluicing phase, all turbines and sluice gates are open, and this phase helps with emptying the basin as quickly as possible. A new holding phase at low water starts when the head difference across the lagoon diminishes. Similar process will follow for the flood tide as shown in Fig. 8.9.

To avoid the influence of water level oscillations during the operation of the turbines and sluice gates, the water levels used to determine the lagoon operation phase and to calculate the discharge through the turbines and sluice gates are calculated by averaging the water level at three points located at 300, 600 and 1000 m away from the lagoon walls.

WSL is represented in the model using domain decomposition, which means that the area inside WSL is simulated as an independent subdomain. Turbines and sluice gates connect the domains inside and outside of the lagoon. A standard orifice equation is used to calculate the discharge of submerged sluice gate as follows:

$$Q = C_d A \sqrt{2gH} \quad (8.11)$$

where H is the water level difference between inside and outside of the lagoon and A is the sluice area. At the preliminary stage of this study, the discharge and the power generated by the turbines were calculated using a Hill chart and discharge coefficient in the orifice equation, namely C_d, which was set to 1 (Baker, 1991).

In addition, a ramp function was used over the effective area, and the discharge through the turbines and sluice gates was implemented in a smooth transition from zero when the turbines and sluice gates were first opened. The ramp function was implemented over a 20-minute period, which avoided generating shock waves to the model as a result of simultaneous operation.

8.4.2 Modelling details

The numerical model TELEMAC is a widely used open source hydro-environmental modelling software system that was developed by the research and development department of Electricite de France (EDF) (https://www.edf.fr/). It contains 2D and 3D free-surface models, called TELEMAC -2D and TELEMAC -3D, which can be coupled with other modules including: the SISYPHE module for sediment transport and morphodynamics modelling, the TOMAWAC module for wave effect modelling, and the WAQTEL module for water quality modelling. Due to the large domain used in this study, and therefore the high simulation time, as well as the well-mixed nature of the estuary, TELEMAC -2D was adopted. The finite-element method is used to solve the governing equations in TELEMAC -2D, using an unstructured triangular mesh, which contains a wide range of numerical options suitable in many fields (Hervouet, 2007). The continuity and the momentum equation of TELEMAC -2D are presented in non-conservative form as follows:

$$\frac{\partial h}{\partial t} + \vec{u} \cdot \overrightarrow{grad}(h) + h\, div(\vec{u}) = Sce \quad (8.12)$$

$$\frac{\partial u}{\partial t}+u\frac{\partial u}{\partial x}+v\frac{\partial u}{\partial y}=-g\frac{\partial Z}{\partial x}+\frac{1}{h}\,div\left(hv_e\overrightarrow{grad}(u)\right)+F_x \tag{8.13}$$

$$\frac{\partial v}{\partial t}+u\frac{\partial v}{\partial x}+v\frac{\partial v}{\partial y}=-g\frac{\partial Z}{\partial y}+\frac{1}{h}\,div\left(hv_e\overrightarrow{grad}(v)\right)+F_y \tag{8.14}$$

where h is the depth of water; Z is the elevation of free surface; u, v are the depth-averaged velocity components in the x- and y-directions; v_e is the effective diffusion, which will model turbulent viscosity and dispersion; Sce is the source term expressed in m/s; F_x, F_y are the source term representing the wind, Coriolis force, bottom friction, a source or a sink of momentum within the domain.

The whole computational domain spans the Severn Estuary and Bristol Channel, which includes an area of 5805 km^2. The open seaward boundary is set at the mouth of the Bristol Channel, which is at a reasonable distance from the scheme and thereby ensures limited impact on the open boundary. A typical spring-neap tidal cycle covers 15.5 days, and the computational domain contains 69,404 nodes and 134,674 elements with the mesh resolution varying from 30 m near the lagoon structure to 600 m close to the open seaward boundary.

The method of characteristics is chosen to solve the advection step. For the discretization in space, a quasi-bubble triangle is applied to the velocity and a linear triangle to the depth. This method of space discretization ensures a balance between accuracy and efficiency. The classic k-ε turbulence model is adopted, together with a Manning's roughness coefficient of 0.025, which is obtained from optimum model calibration. To represent the momentum flux through the turbines or sluice gates, source terms are added to the momentum equations, i.e., Eqs. (8.13) and (8.14) (Ahmadian et al., 2012; Čož et al., 2019). The momentum source in the x direction, induced by the discharge, is represented as follows:

$$F_x^q=\frac{1}{h}(u_{sce}-u)Sce=\frac{1}{h}(u_{sce}-u)\frac{Q}{(\Delta\xi\times\Delta\eta)} \tag{8.15}$$

where $\Delta\xi\times\Delta\eta$ is taken as the area of the source discharge, u is the current velocity at the source point, u_{sce} is the flow velocity through the source point where the turbine or sluice gate is located, u_{sce} is calculated as $\frac{Q}{A}$ where A is the area of the cross-section. A is taken at the cross-section between the turbine blade and the end of the turbine diffusor, as shown in Fig. 8.10.

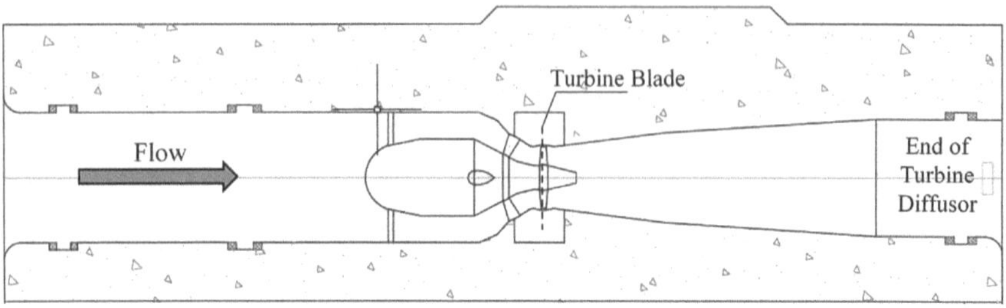

Figure 8.10 Bulb turbine structure and flow cross-section.

Water quality inside the basin could have a significant impact on the secondary benefits of the basin, like those highlighted earlier for Cardiff Bay and likewise for coastal reservoirs in general. However, one of the main concerns in terms of the water quality of a coastal lagoon for tidal energy generation is the water body retention time.

Ahmadian et al. (2010a,b) studied the flushing efficiency of several lagoons, with different turbine formations, along the North Wales coast and found that spreading the turbines more around the periphery of the embankments improves the tidal flushing efficiency. By simulating the transport of a tracer within a tidal stream energy site, Guillou et al. (2019) found that the introduction of turbines in this area has a weak effect on the water renewal capacity, with changes in the residence time being below 5%. The research of Monsen et al. (2002), and da Silva et al. (2019), with regard to the definition of the residence time, focuses on studying the exchange of water between the inside and outside of the basin, and how much time the original water spends inside the impoundment. The volume of water coming into and out of the lagoon, the water jet in the lee of the turbines and sluice gates and the recirculation zone in the lagoon all impact on the dispersion of dissolved pollutants, particulate organic matter and suspended sediments. These influences are reflected in the residence time.

In order to study the residence time of WSL, a mass-conservative passive tracer was considered inside of the lagoon, and then changes in the concentration of tracer inside of lagoon were evaluated. First, the model was run for 5 days to ensure the model is stable hydrodynamically. Then tracer was introduced into the lagoon domain instantaneously and uniformly distributed across the impoundment. Through the research of Guillou et al. (2019), it has also been shown that the residence time depends on the tracer release time during a tidal cycle. For the current study tracer was released inside the lagoon at high water for a typical spring tide.

A similar method was undertaken in the research of Matta et al. (2018), Dabrowski et al. (2012) and Monsen et al. (2002), but different calculation methods of the residence time were applied. For simplicity, the residence time in the current study was defined using the method of Matta et al. (2018) to determine the residence time taken for the tracer concentration in the lagoon to be dropped to 10% of the original concentration.

8.4.3 Model results

The TELEMAC model used in this study was calibrated and validated for an earlier study and could not be calibrated for this study as the lagoon has yet to be built. The details of the prior model validation are not provided here for brevity, but can be found in Abu Bakar et al. (2017) and Guo et al. (2019a,b). The model was then set up to simulate the lagoon operation. This study only uses fixed operation (Xue et al., 2019) and the generation start, and end heads used were 3.5 and 2 m, respectively. Fig. 8.11 shows the water levels inside and outside of the impoundment, flow through the turbines and sluice gates and power generation over the simulation period.

The construction and operation of the lagoon significantly changes the local flow structure around and inside the lagoon. One key where significant changes occur, and therefore where it is crucial to understand the flow structure, is the flow around the sluice gates and turbines (Bray et al., 2016; Čož et al., 2019). Accurate current predictions around and inside the lagoon could contribute to a better understanding of the hydro-environmental/ecological impacts of their operation. However, one of the key challenges in modelling such schemes is accurate modelling of the flow through hydraulic structures and the downstream jets and wakes from through these structures (Bray et al., 2016). To predict these currents accurately then it is essential to ensure conservation of momentum as well as mass (Čož et al., 2019) through turbines and sluice gates. The effects of not including accurate momentum conservation through such structures, where two main approaches to conserve momentum have been considered, are shown in Fig. 8.12. The introduction of a momentum source term has a clear influence on the flow structure near to, and within, a lagoon. As Fig. 8.12.B and C show, the impact of the momentum source term on the flow jet through the turbine depends on the velocity used for momentum adjustment in Eq. (8.15). Fig. 8.12.B is considered to show the most appropriate treatment for momentum conservation as it considers the jet velocity at the end of the diffuser (Čož et al., 2019) as investigated previously with Guo et al. (2019a,b). The scenario without inclusion of a

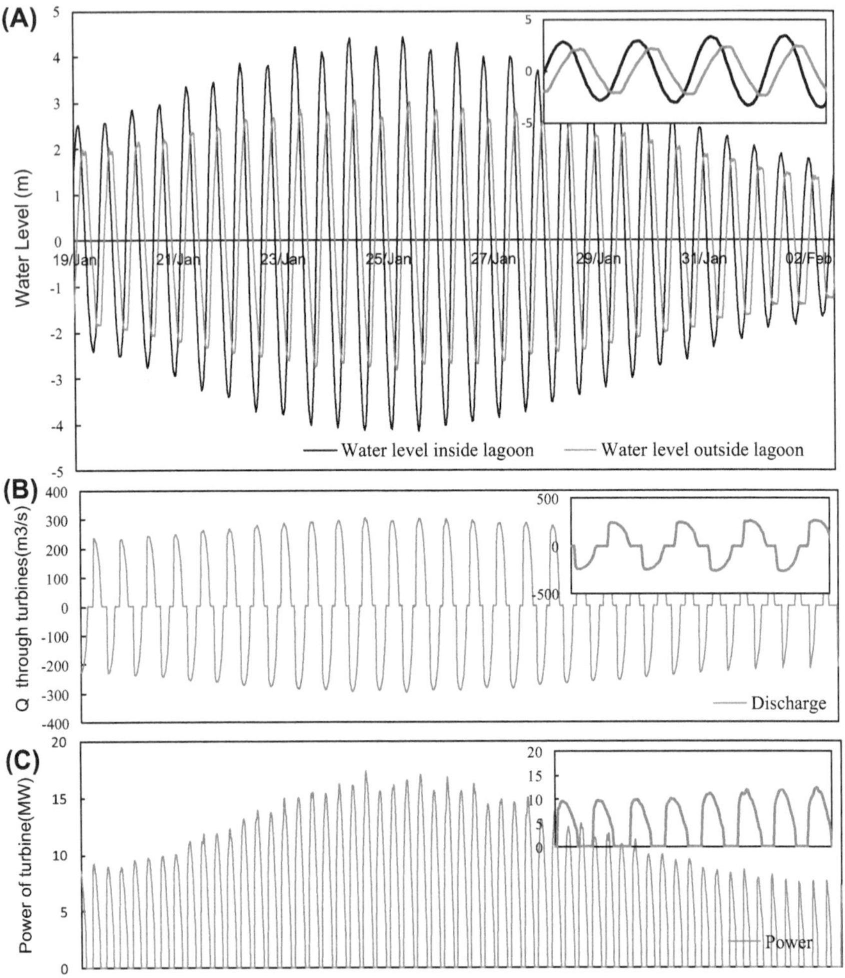

Figure 8.11 Operation and performance of WSL showing: (A) water levels inside and outside of lagoon; (B) discharge through one turbine; and (C) power output from one turbine.

source term for momentum conservation shows the smallest jet and assumes the same velocity for the jet as the receiving water (Hervouet, 2007). This approach therefore introduces different levels of error at different stages of the tide. When adjusting the momentum with a jet velocity taken at the end of the turbine diffusor, most of the water in the wake has a velocity of less than 1 m/s. The high-velocity region, with velocities in the region of 1.5–2 m/s, is increased when the velocity at the blade location is used for the momentum correction. The shape of the wake and the formation of

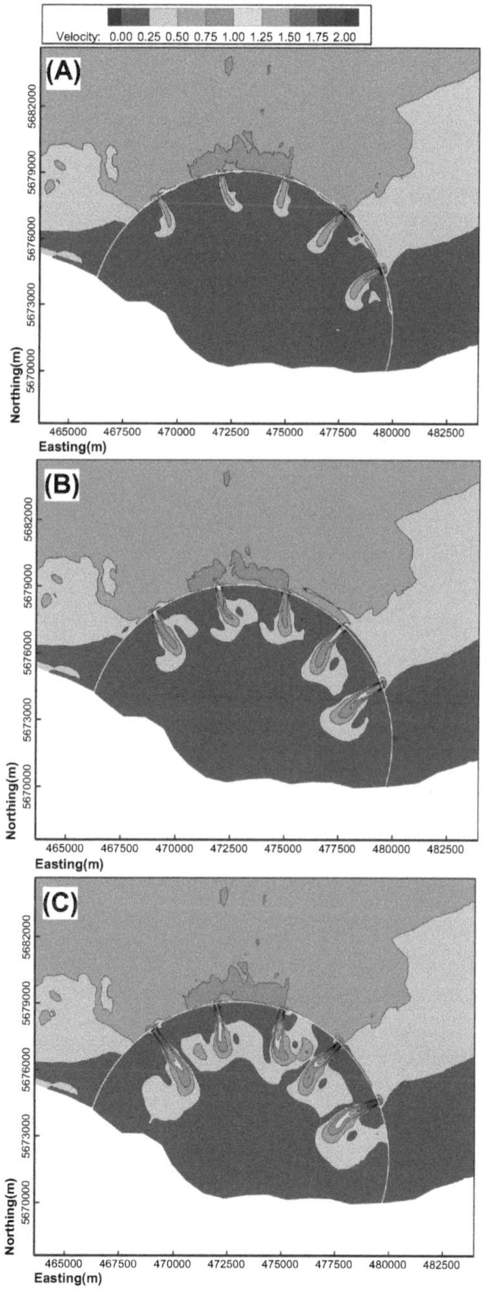

Figure 8.12 Comparison of velocity fields during flood generation for a typical spring tidal cycle: (A) without a momentum source term (i.e., mass conservation only); (B) with source momentum source term with velocity taken at the end of turbine diffusor; (C) with momentum source term with velocity taken at the turbine blades.

eddies are influenced by the shape and size of the basin and the positioning of the turbines and sluice gates as well as velocity of the wakes. This can be seen in comparison with the jets in the basins for other schemes (Falconer et al., 2018; Angeloudis et al., 2016).

The flushing performance of the scheme was also investigated to ensure that water in the basin mixes well with that outside the lagoon. Fig. 8.13

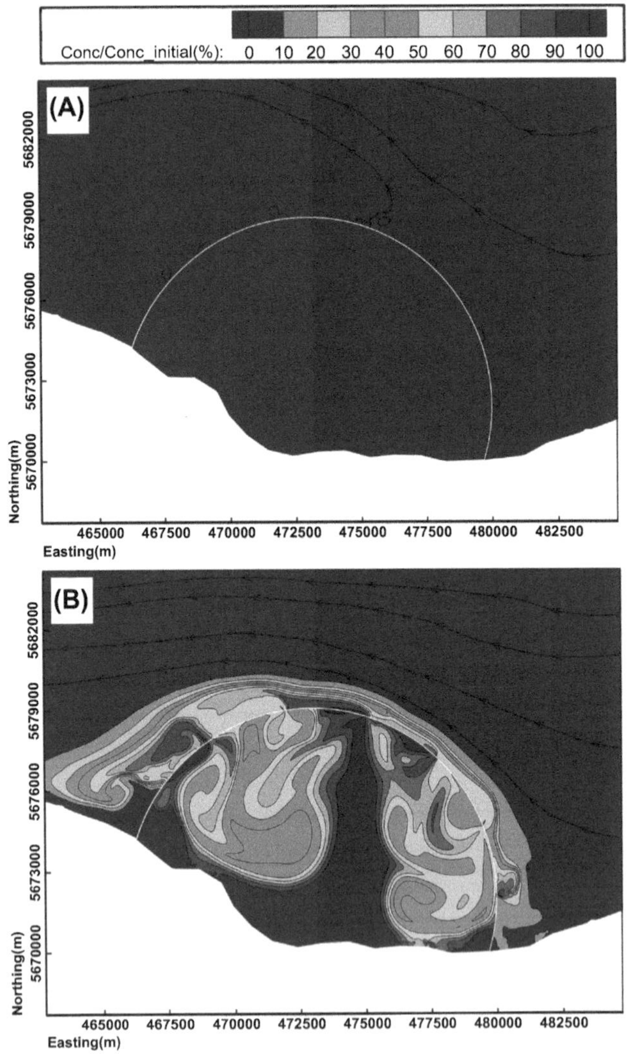

Figure 8.13 Tracer release and flushing process for: (A) release of tracer (at high spring tide), (B) ebb generation phase (11 h after release); (C) flood generation phase (16.5 h after release); and (D) 91.5 h after release.

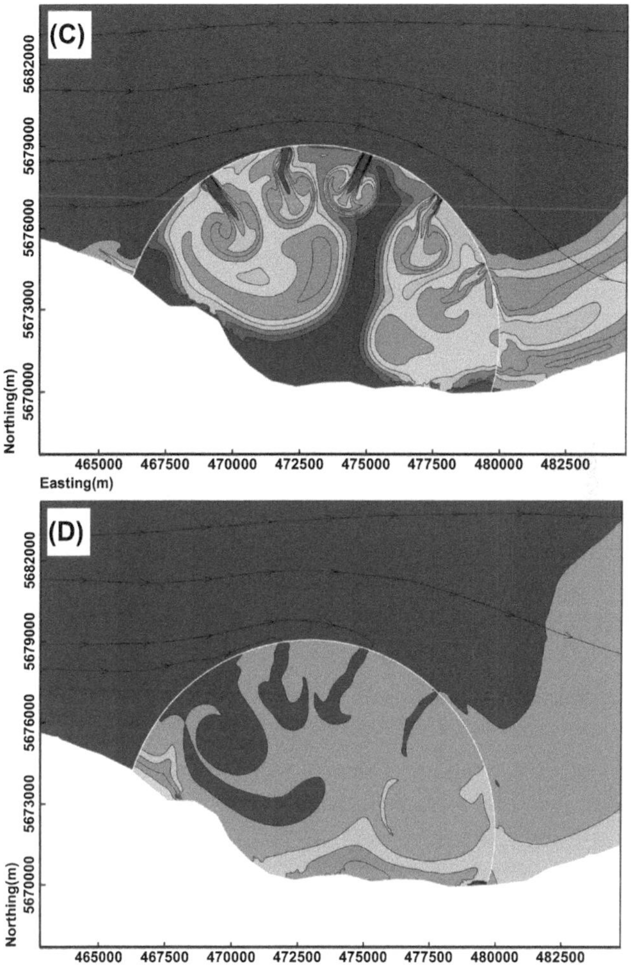

Figure 8.13 Cont'd

shows the tracer distribution and the flushing characteristics after 0, 11, 16.5 and 91.5 h. Flushing of the tracer depends totally on the operation of the lagoon, rather than the natural tide. Also, the turbine wake and the vortices inside the lagoon have a significant impact on the mixing of the tracer with the receiving water.

In order to quantify the mechanisms and timescales of water exchange, the average tracer inside the basin over time has been calculated and shown in Table 8.1. In the scenario of pre-lagoon, the average concentration fluctuates so much in the early stages of the tracer release that there is no

Table 8.1 Average tracer concentration and the time (days) in the lagoon domain.

	$Conc/conc_{initial}$				
Model scenario	**50%**	**40%**	**30%**	**20%**	**10%**
Natural condition			1.30	5.49	15.84
Lagoon without momentum	0.69	1.20	2.22	4.78	13.20
Lagoon with momentum and source velocity taken at the end of diffusor	0.77	1.25	2.00	2.99	8.04
Lagoon with momentum and source velocity taken at the turbine blade	0.76	1.24	1.73	2.64	4.83

structure to limit the tracer movement and the tracer can move freely with the tide. Hence, it is difficult to give an accurate value of the flushing efficiency. Although the tracer concentration decreased rapidly in the beginning, it took much longer (typically 15.84 days) for the tracer to be reduced to 10% of the original value, due to part of the tracer being flushed back to the release zone. The residence time for the model without momentum adjustment and with momentum adjustment, both for the velocities taken at the end of the diffuser and at the blade locations, were 13.20, 8.04 and 4.83 days, respectively. This highlights the importance of accurate representation of momentum conservation in simulating such schemes.

The tracer concentration changes significantly in the lagoon model when comparing the without and with momentum terms, while the difference in using the source velocity at the end of the diffuser, vis-à-vis the blade location, produces similar flushing efficiencies. However, the concentration dropped much faster in the model with the momentum source term based on using the turbine blade source velocity. Thus, the results show that the tracer flushing is directly related to the downstream wake for the turbines and sluice gates. Although the same volume of water comes into the lagoon, the longer downstream wake generates a larger vortex, and this induces a better mixing effect for the fluid within the lagoon. Furthermore, it is noticeable that the residence time is much shorter after construction of a lagoon in comparison with a pre-lagoon scenario. This is thought to be due to the breakwater enabling more estuarine water flowing into the lagoon, resulting in a blockage of the backflow of tracer.

From the tracer flushing research for WSL, it is shown that the operation of the lagoon could enhance the water renewal capacity and contribute to improved pollutant transport in the region, through increased turbulence and increased mixing. Furthermore, by adjusting the lagoon operation procedure, the tracer flushing processes could be accelerated, particularly when emptying the water body inside of lagoon without considering power generation and by keeping the turbines and sluice gates closed to prevent the tracer flowing back into the lagoon.

8.5 Conclusions

This chapter discusses the opportunities for coastal reservoirs to provide urban regeneration and/or tidal energy in coastal and estuarine waters. Whilst most coastal reservoirs having already been built and now operational (such as Qingcaosha Reservoir in the Yangtze Estuary), or currently being planned, are primarily aimed at providing high-quality water at, or near to, the mouth of an estuary for water supply, the design features and the hydro-environmental design characteristics for all coastal reservoirs are similar. Details are therefore first given herein of a general 3D numerical modelling for predicting hydrodynamic, water quality, sediment transport and morphological processes. This is followed by two coastal reservoir examples, including: (1) the Cardiff Bay project where a tidally flushed bay was closed off from the outside estuary to create a freshwater reservoir for urban regeneration; and (2) the West Somerset Lagoon (WSL) for tidal range power generation.

For the Cardiff Bay project, details are given of the application of an integrated modelling tool for predicting the main water quality indicator levels in the coastal reservoir, with particular focus on the faecal coliform levels in the basin from point- and diffuse-source inputs into the two impounded rivers, namely the Taff and Ely. This integrated package consisted of linking a CSO model (namely SWMM) with the HRC's 1D and 3D research models, named FASTER and TRIVAST, respectively. In applying and developing the 3D model, attention was focussed on refining the treatment of bacterial decay, and the model was refined to include decay rates that were dynamically dependent upon sunlight intensity. The model predictions were found to be highly dependent upon the value of the decay rate, and good agreement was obtained between the numerical model predictions and field-measured data. The field data was analysed using a widely used genetic algorithm to provide a new formulation for day- and nighttime decay rates. The model was then applied to the reservoir to assess the variations in the FIO levels across the reservoir and variations with depth, with the results showing relatively high FIO levels in the summer months. Low dissolved oxygen levels were detected in the summer months, and, in a separate study, the DO levels were raised to meet the legal requirement of a minimum DO level of 5 mg/L across the basin using a comprehensive reaeration system.

Telemac-2D has been refined to simulate the operation of the proposed West Somerset Lagoon and its capability to generate electricity. Three different scenarios of the treatment of momentum conservation through the turbines and sluice gates have been studied. The results showed that by including the momentum source term adjustment, to ensure conservation of momentum through the structures, has a significant impact on the downstream wake through the turbines and sluice gates. The residence time was estimated by releasing a tracer in the lagoon basin. The impact of the operation of the lagoon on the residence time was studied by tracking changes in the average tracer concentration within the lagoon impoundment. It was again found that the construction of a lagoon at this site, in the Bristol Channel, could benefit the water renewal capacity in this area, with the tracer flushing processes being related to the inclusion of a momentum source term and its treatment. The residence time was found to decrease with an increase in the size of the downstream wake, due to increased turbulence and vorticity, and better mixing of the sea water with the tracer.

Acknowledgements

The research studies outlined in this chapter were funded by the Natural Environment Research Council and the Engineering and Physical Sciences Research Council. The authors are also grateful to the following organizations for the provision of data: Cardiff Harbour Authority, Natural Resources Wales and the Centre for Research into Environment and Health. The authors are also particularly indebted to colleagues who were involved in the case studies, either directly or indirectly, including Prof. David Kay and Dr Carl Stapleton (CREH), Dr Emma Harris and Dr Seyed Kashefipour.

References

Abu-Bakar, A., Ahmadian, R., Falconer, R.A., 2017. Modelling the transport and decay processes of microbial tracers in a macro-tidal estuary. Water Research 123, 802–824.

Ahmadian, R., Falconer, R., Lin, B., 2010a. Hydro-environmental modelling of proposed Severn barrage, UK. Proceedings of the Institution of Civil Engineers-Energy 163 (3), 107–117.

Ahmadian, R., Morris, C., Falconer, R., 2010b. Hydro-environmental modelling of off-shore and coastally attached impoundments off the north wales coast. In: First IAHR European Congress, Edinburgh (UK), 4–6 May 2010.

Ahmadian, R., Falconer, R.A., Bockelmann-Evans, B., 2014. Comparison of hydro-environmental impacts for ebb-only and two-way generation for a Severn Barrage. Computers and Geosciences 71, 11–19.

Ahmadian, R., Falconer, R., Bockelmann-Evans, B., 2012. Far-field modelling of the hydro-environmental impact of tidal stream turbines. Renewable Energy 38 (1), 107–116.

Angeloudis, A., Falconer, R.A., Bray, S., Ahmadian, R., 2016. Representation and operation of tidal energy impoundments in a coastal hydrodynamic model. Renewable Energy 99, 1103–1115.

Baker, A.C., 1991. Tidal Power. Peter Perengrinus Ltd., London.

Bellair, J.T., Parr-Smit, G.A., Wallis, I.G., 1977. Significance of diurnal variations in faecal coliform die-off rates in the design of ocean outfalls. Journal of Water Pollution Control Federation 2022–2030.

Bray, S., Ahmadian, R., Falconer, R.A., 2016. Impact of representation of hydraulic structures in modelling a Severn barrage. Computers and Geosciences 89, 96–106.

Burt, N., 2002. Cardiff Bay Barrage: overview of hydraulic studies. Proceedings of the ICE Water and Maritime Engineering 154 (2), 93–102.

Cayhono, M., 1993. Three-Dimensional Numerical Modelling of Sediment Transport Processes in Non-stratified Estuarine and Coastal Waters (Ph.D. thesis). University of Bradford, p. 287.

Čož, N., Ahmadian, R., Falconer, R.A., 2019. Implementation of a full momentum conservative approach in modelling flow through tidal structures. Water 11 (9), 1917.

Crompton, D., 2002. Cardiff Bay Barrage. Proceedings of the ICE Water and Maritime Engineering 154 (2), 81–88.

da Silva, D.V., Oleinik, P.H., Costi, J., de Paula Kirinus, E., Marques, W.C., 2019. Residence time patterns of Mirim Lagoon (Brazil) derived from two-dimensional hydrodynamic simulations. Environmental Earth Sciences 78 (5), 163.

Dabrowski, T., Hartnett, M., Olbert, A.I., 2012. Determination of flushing characteristics of the Irish Sea: a spatial approach. Computers and Geosciences 45, 250–260.

Falconer, R.A., Lin, B., 1997. Three-dimensional modelling of water quality in the Humber Estuary. Water Research 31 (5), 1092–1102.

Falconer, R.A., Xia, J., Lin, B., Ahmadian, R., 2009. The Severn Barrage and other tidal energy options: hydrodynamic and power output modelling. Science in China Series E: Technological Sciences 52 (11), 3413–3424.

Falconer, R.A., Angeloudis, A., Ahmadian, R., 2018. Modelling Hydro-Environmental Impacts of Tidal Range Renewable Energy Projects in Coastal Waters. Handbook of Coastal and Ocean Engineering, pp. 1553–1574.

Gameson, A.L.H., Gould, D.J., 1975. Effects of solar radiation on the mortality of some terrestrial bacteria in sea water. In: Gameson, A.L.H. (Ed.), Discharge of Sewage from Sea Outfalls. Pergamon Press, Oxford, pp. 209–217.

Gameson, A.L.H., Saxon, J.R., 1967. Field studies on effect of daylight on mortality of coliform bacteria. Water Research 1, 279–295.

Guillou, N., Thiébot, J., Chaplain, G., 2019. Turbines' effects on water renewal within a marine tidal stream energy site. Energy 189, 116113.

Guo, B., Ahmadian, R., Falconer, R.A., 2019a. Studying the wake of an island in macro-tidal estuary. Estuaries and Coasts (Under review).

Guo, B., Ahmadian, R., Falconer, R.A., 2019b. Influence of different momentum sources on modelling tidal lagoons. In: Proceedings of the 38th IAHR World Congress, Panama City.

Harris, E.L., 2003. Environmental Hydroinformatics Tools for Recreational Water Quality Modelling (Ph.D. thesis). University of Wales, Cardiff, UK, p. 208.

Harris, E.L., Falconer, R.A., Kay, D., Stapleton, C., 2002. Development of a modelling tool to quantify faecal indicator levels in Cardiff Bay. Proceedings of the ICE Water and Maritime Engineering 154 (2), 129–135.

Hervouet, J.M., 2007. Hydrodynamics of free surface flows: modelling with the finite element method. Wiley.

Hunter, P., Gander, H.C.W., 2002. Cardiff Bay Barrage: planning and design. Proceedings of the ICE Water and Maritime Engineering 154 (2), 117–128.

Kashefipour, S.M., Lin, B., Harris, E.L., Falconer, R.A., 2002. Hydro-environmental modelling for bathing water compliance of an estuarine basin. Water Research 36 (7), 1854–1868.

Kay, D., Stapleton, C.M., Wyer, M.D., McDonald, A.T., Crowther, J., Paul, N., Jones, K., Francis, C., Watkins, J., Humphrey, N., Lin, B., Yang, L., Falconer, R.A., Gardner, S., 2005. Decay of intestinal enterococci concentrations in high energy estuarine and coastal waters: towards real-time T90 values for modelling faecal indicators in recreational waters. Water Research 39, 655–667.

Leonard, B.P., 1991. The ULTIMATE conservative difference scheme applied to unsteady one-dimensional advection. Computer Methods in Applied Mechanics and Engineering 88, 17–74.

Lin, B., Falconer, R.A., 1997. Three-dimensional layer integrated modelling of estuarine flows with flooding and drying. Estuarine, Coastal and Shelf Science 44, 737–751.

Lin, B., Kashefipour, S.M., Falconer, R.A., 2003. Predicting nearshore coliform bacteria using ANNs. Water Science and Technology 48 (10), 225–232.

Matta, E., Koch, H., Selge, F., Simshäuser, M.N., Rossiter, K., da Silva, G.M.N., Gunkel, G., Hinkelmann, R., 2018. Modeling the impacts of climate extremes and multiple water uses to support water management in the Icó-Mandantes Bay, Northeast Brazil. Journal of Water and Climate Change 10 (4), 893–906.

Monsen, N.E., Cloern, J.E., Lucas, L.V., Monismith, S.G., 2002. A comment on the use of flushing time, residence time, and age as transport time scales. Limnology and Oceanography 47 (5), 1545–1553.

Stapleton, C., Wyer, M., Crowther, J., Francis, F., Watkins, J., Rushby, L., Wilkinson, J., Kay, D., 2002. Towards dynamical modelling of bacterial concentrations in the near-shore zone: I Driving the models with dynamic input variables and assumptions. Proceedings of the 13th International Association for Hydraulic Research/APD Congress Singapore. World Scientific Vol II August 1025–1029.

Xia, J., Falconer, R.A., Lin, B., 2010. Impact of different tidal renewable energy projects on the hydrodynamic processes in the Severn Estuary, UK. Ocean Modelling 32 (1–2), 86–104.

Xue, J., Ahmadian, R., Falconer, R.A., 2019. Optimising the operation of tidal range schemes. Energies 12 (15), 2870.

CHAPTER 9

Impacts of climate change on coastal infrastructure

Simon Beecham
Natural and Built Environments Research Centre, University of South Australia, Adelaide, SA, Australia

9.1 Introduction

Climate change will bring about significant changes to the capacity of, and the demand on, water resources. The resulting changes include increasing climate variability that is expected to affect hydrologic conditions. The effects of climate variability on meteorological variables have been extensively observed in many regions around the world. Of these, rainfall is one of the most important variables. Understanding the effects of climate variability on spatial and temporal rainfall characteristics is of particular interest to water resource policy makers. Investigating rainfall variability at the regional scale is essential for understanding potential impacts on both human communities and the natural environment (Beecham and Chowdhury, 2012).

An example of a measurable change in atmospheric circulation has occurred over the last 50 years in Australia. During this time, rain-bearing systems in the southwest region of Western Australia (SWWA) have moved southwards. The effects of this change include a later commencement of winter rains, reduced winter rainfall and lower overall rainfall intensities. Perth, which is Australia's fourth most populous city, receives an average annual rainfall of approximately 870 mm. The broader catchments in the SWWA, which include Perth, have experienced a significant decline in average rainfall since the 1970s. Such a steady and consistent decline cannot be due either to short-term droughts or even to long-term decadal-scale oscillations. Instead, the decline, which has been studied extensively, has been associated with a reduced number of westerly frontal systems and low-pressure troughs combined with a reduction in the rainfall volume derived from rain-bearing systems (Smith and Power, 2014). These climatic changes have greatly reduced the inflows into Perth's 11 major dams (Fig. 9.1), which have been measured since 1911. The annual average inflow reduced from approximately 350 GL in the middle of the 20th century to

Sustainable Water Resource Development Using Coastal Reservoirs
ISBN 978-0-12-818002-0
https://doi.org/10.1016/B978-0-12-818002-0.00009-5

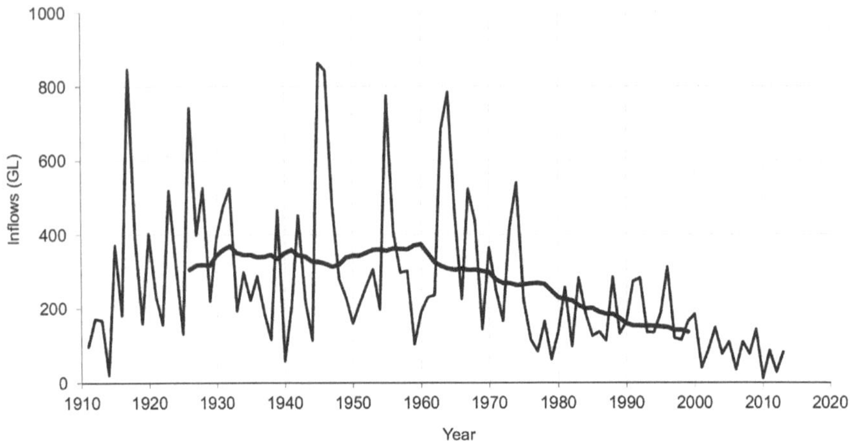

Figure 9.1 Annual inflows to Perth Dams together with 31-year running averages *(After Smith, I., Power, S., 2014. Past and future changes to inflows into Perth (Western Australia) dams, Journal of Hydrology: Regional Studies 2, 84–96.)*

around 175 GL by the year 2000. The average inflow was recorded at its lowest value in 2010 when only 12 GL flowed into the 11 dams.

Driving this decline in dam inflows has been a decrease in annual rainfall and an increase in average annual temperature. During the period 1911–1965, 20% of the recorded annual rainfalls exceeded 800 mm. After 1965, no annual rainfalls were greater than 800 mm and the lowest amount recorded was 400 mm in 2010. Over the same period, annual mean temperatures showed a positive trend, increasing by approximately +0.8°C per 100 years with the hottest three years being the last three on record (2011, 2012 and 2013) for the Smith and Powell study.

Coastal reservoirs would be difficult to implement in this region because of the high permeability of the inland soils. Instead, the State Government response to this decrease in water supplies has been to turn to seawater desalination as this is viewed as an unlimited resource that is not climate-dependent. Perversely though, the high energy consumption associated with seawater desalination could further fuel global warming through additional greenhouse gas emissions.

9.2 Influence of climate indices

Rainfall is influenced by various natural climate phenomena. In Australia, these key climate drivers largely originate in the Indian, Pacific and Southern Oceans. They include the El Niño–Southern Oscillation (ENSO), the Pacific and Indian Ocean sea surface temperature (SST) variability, the subtropical ridge, the Southern Annular Mode and the Madden–Julian Oscillation (Chowdhury and Beecham, 2015). ENSO can be measured by a number of indicators including the Southern Oscillation Index (SOI) and the Niño3.4 index. SOI is calculated from the sea level air pressure difference between Tahiti and Darwin. The Niño3.4 index is the average of SST anomalies over the region 5°N–5°S and 170°W–120°W.

The Indian Ocean Dipole (IOD) is a coupled ocean-atmosphere phenomenon in the Indian Ocean (Saji et al., 1999) that is measured by the Dipole Mode Index (DMI), which is the difference between SST in the western (50°E–70°E and 10°S–10°N) and eastern (90°E–110°E and 10°S–0°S) equatorial Indian Ocean.

The ENSO and IOD phenomena significantly influence eastern and southern Australian rainfall, respectively (Cai et al., 2011). For South American summer rainfall, ENSO is the dominant source of variability (Grimm, 2011). In the Pearl River basin in China, rainfall variability is dominated by the IOD (Niu, 2012). In the Colorado River Basin in the United States, both ENSO and the Pacific Decadal Oscillation affect rainfall (Kim et al., 2008).

To illustrate the temporal variation in the influence of climate indices on rainfall in one region of Australia (South Australia), Fig. 9.2 shows the correlation with winter rainfall for SOI, DMI and Niño3.4, while Fig. 9.3 shows the corresponding correlations for spring rainfall. In both figures, a small red circle (grey circle in printed version) indicates statistically significant correlation at the 5% significance level between rainfall and the three climate drivers. A large red circle (grey circle in printed version) indicates statistically significant correlation at the 1% significance level.

The background rainfall images in Fig. 9.2 and Fig. 9.3 show 1961–90 average rainfall in mm. From Fig. 9.2, winter rainfall in the south and east parts of South Australia is strongly influenced by both SOI and DMI, particularly in July and August. Both SOI and DMI are intercorrelated in winter. From Fig. 9.3, spring rainfall is significantly influenced by DMI in the south and east parts of South Australia, particularly in September and October. In terms of ENSO phenomena, whilst both SOI and Niño3.4 are correlated, SOI appears to be more influential than Niño3.4 for South Australia winter and spring rainfall.

In terms of coastal reservoir design, these results indicate clearly the importance of understanding the temporal and spatial influences of climate drivers for the relevant catchment. It is then important to select good performing general circulation models (GCMs) that can reproduce these influences in their climate projections.

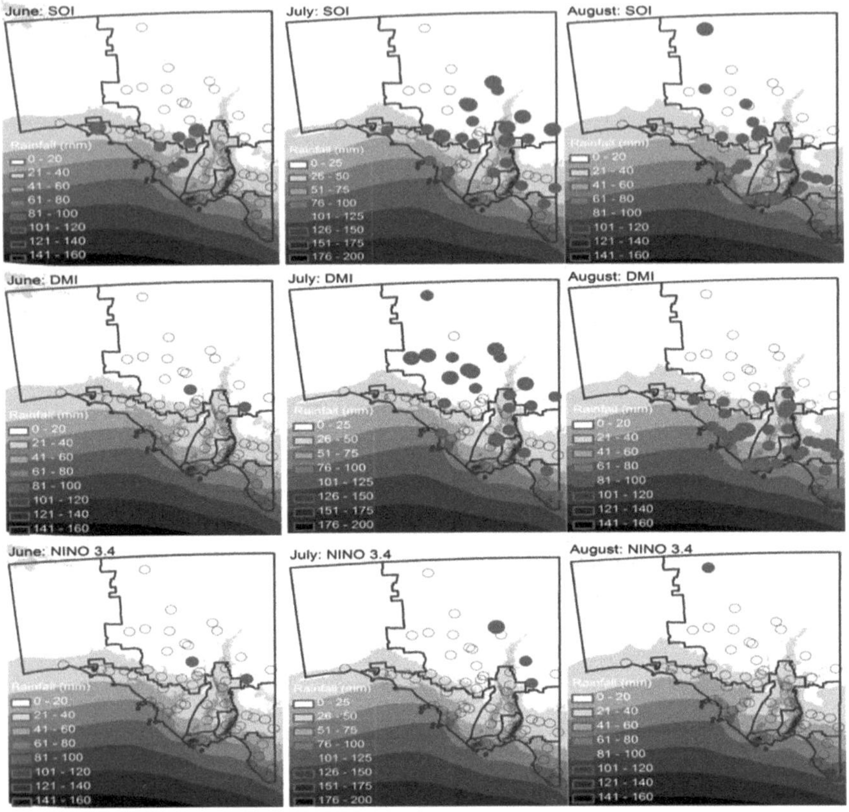

Figure 9.2 Impact of the El Niño–Southern Oscillation (ENSO) and Indian Ocean Dipole (IOD) on winter rainfall in South Australia. *(After Chowdhury, R. K., Beecham, S., 2013. Influence of SOI, DMI and Niño3.4 on south Australian rainfall, Stochastic Environmental Research and Risk Assessment 27(8), 1909–1920.)*

9.3 Climate modelling

Atmospheric circulation, topography, land use and other regional features modify global climate changes to produce unique patterns of change at the regional scale. As the future changes to these water resources cannot be measured in the present, hydrological models are critical in the planning required to adapt our water resource management strategies to future climate conditions. Such models include catchment runoff models, reservoir management models, flood prediction models, water quality models, groundwater recharge and flow models and crop water balance models.

Estimating regional climate change impacts on coastal reservoir systems involves two domains, namely the climate system and the coastal reservoir

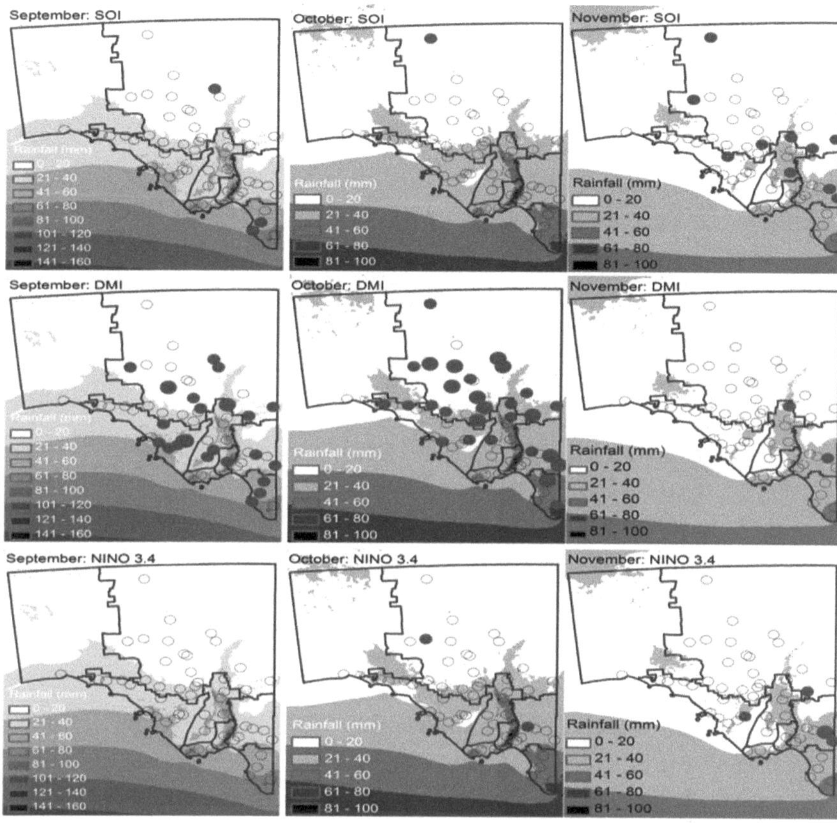

Figure 9.3 Impact of the El Niño–Southern Oscillation (ENSO) and Indian Ocean Dipole (IOD) on spring rainfall in South Australia. *(After Chowdhury, R. K., Beecham, S., 2013. Influence of SOI, DMI and Niño3.4 on south Australian rainfall, Stochastic Environmental Research and Risk Assessment 27(8), 1909–1920.)*

system, as shown in Fig. 9.4. In terms of the climate system, GCMs can model the effects of anthropogenic climate forcing caused by different greenhouse gas concentrations in the atmosphere. Different scales of climate models, including global (GCMs) and/or regional climate models (RCMs), are available. Because of structural differences in the models, climate projections differ between individual GCMs. It is therefore common practice to use ensemble GCM projections, which are downscaled and then corrected for bias.

In terms of a coastal reservoir system, various deterministic and/or probabilistic simulation models are available which can transform the downscaled rainfall inputs into stream flows.

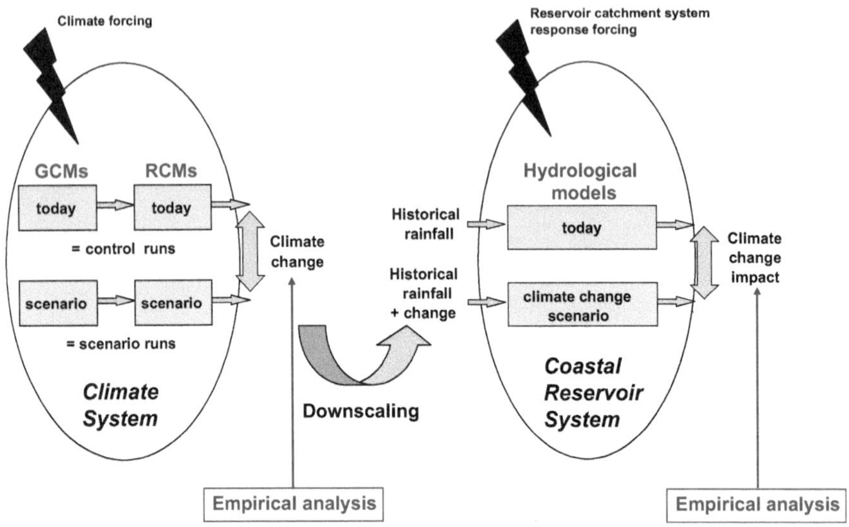

Figure 9.4 Hydrological impact analysis of climate change for a coastal reservoir system. *GCMs*, general circulation models; *RCMs*, regional climate models. *(Adapted from Willems, P., Olsson, J., Arnbjerg-Nielsen, K., Beecham, S., Pathirana, A., Gregersen, I. B., Madsen, H., Nguyen, V.-T.-V., 2012. Impacts of Climate Change on Rainfall Extremes and Urban Drainage Systems, IWA Publishing, England.)*

The GCM and/or RCM outputs can be validated based on recorded observations, including by analyzing trends in long-term rainfall records. In Fig. 9.4, this is denoted *Empirical analysis*. For future scenarios, rainfall and other hydrological changes simulated in the climate system output become inputs to the coastal reservoir hydrological models.

Owing to their large spatial and temporal resolutions, the outputs from GCMs and RCMs require downscaling. Dynamic downscaling refers to using a nested RCM to downscale GCM projections. A more common technique is to use statistical downscaling, which uses statistical relationships to relate large-scale climate variables to the local-scale climate.

Uncertainties are introduced throughout the processes shown in Fig. 9.4. These uncertainties include the choice of a climate forcing scenario, the natural variability of the climate and catchment and the parameterization of the various models.

9.4 Statistical downscaling for a coastal catchment

To illustrate how climate modelling might be applied to a catchment containing a proposed coastal reservoir, the Onkaparinga catchment in South Australia (Fig. 9.5) is now considered as a hypothetical case study.

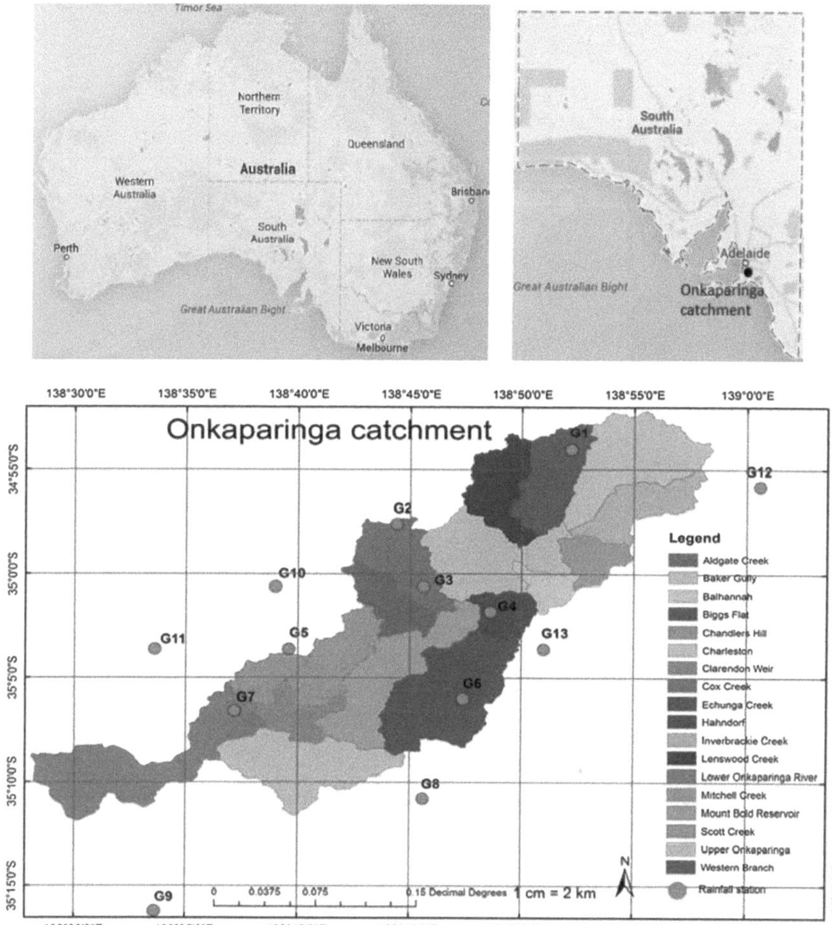

Figure 9.5 Subcatchments and rainfall stations of the Onkaparinga catchment. *(Adapted from Rashid, M. M., Beecham, S., Chowdhury, R. K., 2015a. Assessment of trends in point rainfall using Continuous Wavelet Transforms, Advances in Water Resources 82, 1–15.)*

The Onkaparinga catchment is 560 km^2 in area. It is located 26 km south-east of the metropolitan area of Adelaide and provides up to 60% of that city's municipal water as well as providing much of the local area's irrigation water. Rashid et al. (2015b) found the rainfall in the Onkaparinga catchment to have a large spatial variability and this makes it a useful case study for climate downscaling. As shown in Fig. 9.6, even at 560 km^2 in size, the catchment appears small when plotted in the 2.5 × 2.5 degrees NCEP/NCAR Reanalysis data set grid. The NCEP/NCAR Reanalysis data set is a

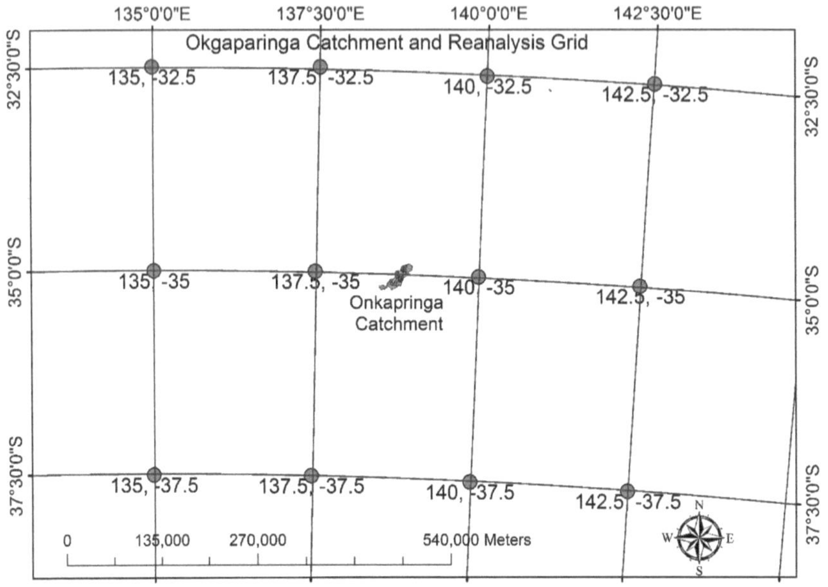

Figure 9.6 National Centers for Environmental Prediction/National Center for Atmospheric Research reanalysis grid points (2.5 × 2.5 degrees) around the Onkaparinga catchment. *(Adapted from Rashid, M. M., Beecham, S., and Chowdhury, R. K., 2015c. Statistical downscaling of CMIP5 outputs for projecting future changes in rainfall in the Onkaparinga catchment, The Science of the Total Environment 530–531, 171–182.)*

continually updated (1948 to present) globally gridded data set that represents the state of the Earth's atmosphere, incorporating observations and numerical weather prediction model outputs from 1948 to present. It is a joint product from the National Centers for Environmental Prediction (NCEP) and the National Center for Atmospheric Research (NCAR). This observed gridded data are often used in the regression of GCM outputs.

The methodology for generating catchment scale projections of future streamflows in a catchment that has large rainfall variability is as follows:

1. An ensemble of Coupled Model Intercomparison Project 5 (CMIP5) GCMs is selected. Suitable GCMs for this catchment might be CSIRO-MK3.6.0 by the Commonwealth Scientific and Industrial Research Organization in collaboration with the Queensland Climate Change Centre of Excellence, GFDL-ESM2M by the NOAA Geophysical Fluid Dynamics Laboratory and MIROC-ESM by the Japan Agency for Marine-Earth Science and Technology. Historical daily outputs of these GCMs for the period 1961–2000 and future daily outputs for the period 2041–60 for two Representative Concentration

Pathway (RCP) scenarios, namely RCP4.5 and RCP8.5, can be collated from the Earth System Grid data distribution portal (www.earthsystemgrid.org). NCEP reanalysis data can be extracted from the 12 grid points around the Onkaparinga catchment shown in Fig. 9.6. The GCM predictor outputs will need to be linearly interpolated to align with the NCEP reanalysis grid resolution.

2. A statistical downscaling model is selected. The Generalized Linear Modelling of daily Climate sequence (GLIMCLIM) is a suitable multisite stochastic downscaling model (Chandler, 2002) based on a generalized linear model. GLIMCLIM can be used to statistically downscale daily rainfall at the rainfall stations in the Onkaparinga catchment shown in Fig. 9.5. The model can be calibrated using predictors from the NCEP reanalysis datasets over the period 1961–86. Large-scale circulation and atmospheric variables can be selected from the NCEP reanalysis datasets and these need to be correlated with observed rainfall. Usually 10 variables are sufficient to be used as predictors in the model. Each predictor variable is extracted and averaged from the 12 reanalysis grid points (2.5 × 2.5 degrees) around the study area (Fig. 9.6). Details of the predictor selection process are reported in Beecham et al. (2014). The model can be validated over the period 1987–2000. Once the model is validated, it can be used to project the future daily rainfall for different GCMs and climate forcing scenarios for future periods. Moreover, the model can be run with historical outputs from the GCMs to compare how well the model simulates rainfall for the calibration and validation periods. Any bias can be corrected using a Frequency Adapted Quantile Mapping bias correction technique. Finally, the future rainfall projections from the GLIMCLIM model can be used for calibration and validation of a rainfall–streamflow model.
3. A rainfall–streamflow model can be developed, in which the downscaled rainfall series described in step 2 above and its wavelet-based decomposed subseries at optimum lags are used as covariates in a generalized additive model for location, scale and shape. Details of such a rainfall–streamflow model, including the wavelet decomposition process, are presented in Rashid and Beecham (2019).

Using the climate modelling approach described above, Rashid and Beecham (2019) found there to be a significant reduction in streamflow in three of the largest subcatchments in the Onkaparinga catchment (Fig. 9.7).

This result is consistent with the findings of Westra et al. (2015). From Fig. 9.7, a greater reduction in streamflow will occur in spring and summer

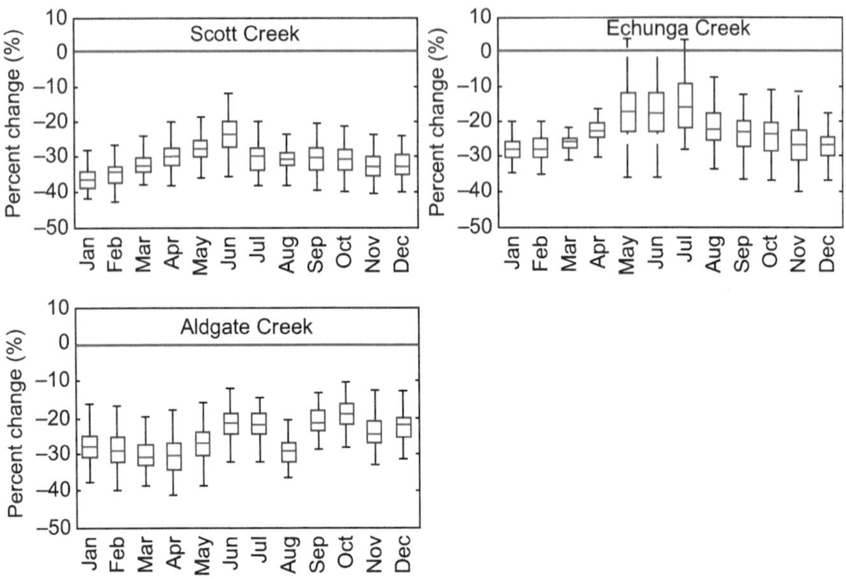

Figure 9.7 Percentile changes in projected streamflow for the RCP4.5 scenario for the period 2041–60 compared with the reference period 1973–2000. The boxplots depict the range of median and interquartile range (IQR) values for each model for 100 simulations; the whiskers represent 1.5 IQR from the box end. *(After Rashid, M. M., Beecham, S., 2019. Simulation of streamflow with statistically downscaled daily rainfall using a hybrid of wavelet and GAMLSS models, Hydrological Sciences Journal 64 (11), 1327–1339.)*

than in winter. This is also consistent with the projected pattern of future reductions in rainfall (Rashid et al., 2015c) for the Onkaparinga catchment. This demonstrates that the rainfall–streamflow model can accurately project future changes in streamflow.

A further important consideration is climate change impacts on water quality. Yang et al. (2013) discussed how, in the context of coastal reservoir systems, surface waters are more susceptible to pollution than groundwater. This is because of the difficulty of protecting surface waters from contamination from upstream surface runoff. They also reported that water quality was not as consistent as for groundwater and that in particular temperature and turbidity levels fluctuate because of their dependence on the amount of rainfall. Soil erosion and the related siltation of coastal reservoirs are also important water quality concerns. Yang et al. (2013) reported that the storage capacity of the world's on-land reservoirs is decreased by more than 1% annually due to soil erosion and siltation. They also related this to climate change and stated that reservoir sedimentation rates would be accelerated because of the severity of storm events as a result of global warming.

9.5 Conclusions

In many countries, water resource systems are particularly vulnerable to rapid population growth and climate change. In the presence of climate change–induced uncertainty, water resource systems need to be more resilient and multisourced. Decreasing volumetric rainfall trends impact on reservoir yields and operational practices. Policy makers, end users and researchers need to work together to develop a consistent approach to interpreting the effects of climate variability and change on water resources.

Coastal reservoir systems are particularly susceptible to climate change because they are located at or very close to sea level. Any changes to sea level from global warming or changes to either fluvial flows or sea surge patterns because of climate change will potentially impact coastal reservoir systems.

References

Beecham, S., Chowdhury, R.K., 2012. Effects of changing rainfall patterns on WSUD in Australia. Proceedings - Institution of Civil Engineers: Water Management 165 (5), 285–298.

Beecham, S., Rashid, M.M., Chowdhury, R.K., 2014. Statistical downscaling of multi-site daily rainfall in a South Australian catchment using a generalized linear model. International Journal of Climatology 34 (14), 3654–3670.

Cai, W., Rensch, P.V., Cowan, T., Hendon, H.H., 2011. Teleconnection pathways of ENSO and the IOD and the mechanisms for impacts on Australian rainfall. Journal of Climate 24, 3910–3923.

Chandler, R.E., 2002. GLIMCLIM: Generalised Linear Modelling for Daily Climate Time Series (Software and User Guide), Research Report No. 227. Department of Statistical Science, University College London.

Chowdhury, R.K., Beecham, S., 2013. Influence of SOI, DMI and Niño3.4 on south Australian rainfall. Stochastic Environmental Research and Risk Assessment 27 (8), 1909–1920.

Chowdhury, R.K., Beecham, S., Boland, J., Piantadosi, J., 2015. Understanding South Australian rainfall trends and step changes. International Journal of Climatology 35 (3), 348–360.

Grimm, A.M., 2011. Interannual climate variability in South America: impacts on seasonal precipitation, extreme events, and possible effects of climate change. Stochastic Environmental Research and Risk Assessment 25 (4), 537–554.

Kim, T.W., Yoo, C., Ahn, J.H., 2008. Influence of climate variation on seasonal precipitation in the Colorado River Basin. Stochastic Environmental Research and Risk Assessment 22 (3), 411–420.

Niu, J., 2012. Precipitation in the Pearl River basin, South China: scaling, regional patterns, and influence of large-scale climate anomalies. Stochastic Environmental Research and Risk Assessment 27 (5), 1253–1268.

Rashid, M.M., Beecham, S., Chowdhury, R.K., 2015a. Assessment of trends in point rainfall using Continuous Wavelet Transforms. Advances in Water Resources 82, 1–15.

Rashid, M.M., Beecham, S., Chowdhury, R.K., 2015b. Statistical characteristics of rainfall in the Onkaparinga catchment in South Australia. Journal of Water and Climate Change 6 (2), 352–373.
Rashid, M.M., Beecham, S., Chowdhury, R.K., 2015c. Statistical downscaling of CMIP5 outputs for projecting future changes in rainfall in the Onkaparinga catchment. The Science of the Total Environment 530–531, 171–182.
Rashid, M.M., Beecham, S., 2019. Simulation of streamflow with statistically downscaled daily rainfall using a hybrid of wavelet and GAMLSS models. Hydrological Sciences Journal 64 (11), 1327–1339.
Saji, N.H., Goswami, B.N., Vinayachandran, P.N., Yamagata, T., 1999. A dipole mode in the tropical Indian Ocean. Nature 401, 360–363.
Smith, I., Power, S., 2014. Past and future changes to inflows into Perth (Western Australia) dams. Journal of Hydrology: Regional Studies 2, 84–96.
Westra, S., Thyer, M., Leonard, M., Kavetski, D., Lambert, M., 2015. Impacts of Climate Change on Surface Water in the Onkaparinga Catchment. Final Report Volume 1: Hydrological Model Development and Sources of Uncertainty. Goyder Institute for Water Research, Adelaide, South Australia. Technical Report Series No. 14/22. ISSN: 1839-2725.
Willems, P., Olsson, J., Arnbjerg-Nielsen, K., Beecham, S., Pathirana, A., Gregersen, I.B., Madsen, H., Nguyen, V.-T.-V., 2012. Impacts of Climate Change on Rainfall Extremes and Urban Drainage Systems. IWA Publishing, England.
Yang, S., Liu, J., Lin, P., Jiang, C., 2013. Coastal reservoir strategy and its applications. Chapter 5. In: Wurbs, R. (Ed.), Water Resources Planning, Development and Management. Intech, United States, pp. 95–115.

CHAPTER 10

Challenges and opportunities for coastal reservoir development in India

T.G. Sitharam[1,2], Sreevalsa Kolathayar[3]
[1]Department of Civil Engineering, Indian Institute of Science, Bangalore, Karnataka, India; [2]IIT Guwahati, Guwahati, Assam, India; [3]Department of Civil Engineering, National Institute of Technology Karnataka, Surathkal, Karnataka, India

10.1 Introduction

Demand for water keeps on increasing day by day; however, average rainfall has remained constant and it is plentiful. In India, average annual rainfall is 1172 mm and the total volume amounts to 4000 billion cubic metres (BCM). Most of India's rainfall comes over a 4-month period starting from June to September. India also experiences years of excess monsoons and floods, followed by below average or late monsoons with droughts. Despite abundant rains in June to September, some regions face shortage of drinking water, while some other parts of the country receive excessive rains resulting in floods. This geographical and time variance in availability of natural water versus the year-round demand for irrigation, drinking and industrial water creates a demand-supply gap that keeps worsening with rising population. It is estimated that about 4400 Thousand Million Cubic Feet (TMC) ft (125 BCM) of river floodwater from west flowing rivers alone drains into Arabian Sea. From this it is very clear that India is not water-starved but it is storage-starved. Per capita storage of water in India for domestic purpose is too small when compared with developed countries. The currently available water solutions are unable to satisfy people's increasing need for water. As far as rainwater is concerned, many regions in the country receive too much rain in the wet season, thereby causing floods and excess rainwater discharge into the sea, while receiving a shortage of rainfall in dry season. Hence, the solution lies in utilizing or conserving the abundant monsoon water which runs off to the ocean. It is wise to invest in storage of excess river floodwater (which is reasonably clean) in the coastal reservoirs close to the place where river joins the sea.

Sustainable Water Resource Development Using Coastal Reservoirs
ISBN 978-0-12-818002-0
https://doi.org/10.1016/B978-0-12-818002-0.00010-1

There is a need for a method to develop a sustainable water source for managing excess floodwater by creating coastal reservoirs at appropriate locations along the coast. The floodwater has to be stored in coastal reservoirs by building impermeable seawalls. The primary objective of this chapter is to review the opportunities for developing a coastal reservoir in India as a sustainable water source by managing rainwater and river floodwaters. The challenges and issues for creating such a potential coastal reservoir for storing river floodwater are also presented.

10.2 Water supply options

There are different kinds of water supply options available in the world, for example, groundwater, inland reservoir, river water, desalination of seawater, reuse of wastewater and diversion of water from a remote source (interbasin water transfer), etc. Each of these has their own merits and limitations. The primary renewable source of freshwater is rainfall. Surface water, subsurface flow, groundwater and seawater are other sources of water available for water supply options. As the available freshwater sources are very less throughout the world, there is a stress upon the available sources and conservation structures. Focus on desalination and other techniques is increased. However, for a developing country like India, it is very difficult to make a huge investment in projects like desalination, which requires a high maintenance cost and thus it is not viable. Desalination is not environmental-friendly as well as it affects marine life where brine is released. The coastal cities are normally more urbanized and more and more urbanization is concentrated along the coasts. There is a rapid migration of people to coastal cities and towns because of the job opportunities. There would be rapid increase in population and the demand for water would increase exponentially. Presently, the solution for this issue would be a coastal reservoir, i.e., a reservoir in the sea to store a portion of the river floodwaters, which is joining the ocean in the monsoon period, and use it during the drought periods (Kolathayar et al., 2019).

10.3 Coastal reservoir to store floodwater

Coastal reservoir is a unique structure constructed at an estuary or gulf or bay or in the sea (Point where a river meets a sea) to store the portion of excess water at flood times. This seawall or dike structure may run for kilometres together in the coastline. Coastal reservoir has many advantages such as no land acquisition problems, no land submergence and forest submergence like in inland reservoirs. They are already put to work in countries such as the Netherlands, Singapore, China, India, South Korea, Hong Kong, United Kingdom, etc., and are proven to be beneficial. They are designed with gates so that the excess floodwater, which is more than the capacity of the reservoir, can be discharged into the sea. Even inlet and outlets are controlled such that only good quality water enters the reservoir with very little stagnation of water (Yang, 2010; Yang, 2015; Yang and Kelly, 2015; Yang et al., 2005; Yang et al., 2013). One can adopt very innovative designs with smart sensor network to efficiently operate the coastal reservoirs.

Fig. 10.1 shows a typical schematic of coastal reservoir which enables storing of freshwater during river floods. The freshwater reservoir includes an impermeable seawall for containing the freshwater and preventing a mixing of the freshwater with the seawater. The seawall is located on the sea bed having three-sided structure to prevent ingress of seawater into the freshwater/river course. By keeping out seawater, by construction of seawall along with one or two barrage (with sluice gates) close to the mouth of river forms the freshwater reservoir in the ocean. Sea-based reservoir is part of the river watershed which is formed by the convergence of the rivers and allows freshwater and salts to join the ocean. The reservoir will be

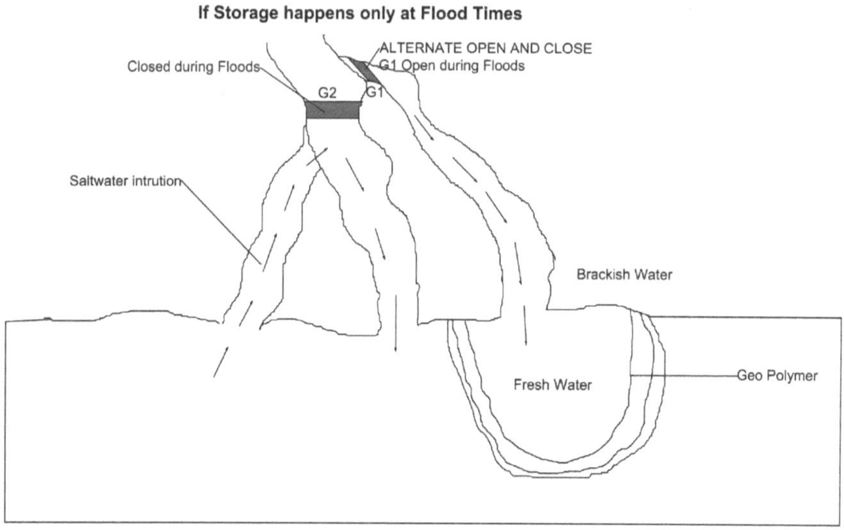

Figure 10.1 Schematic of a coastal reservoir that enables storage of freshwater during river floods.

shallow waters and have an average depth of 10–20 m with a maximum depth of about 30 m. When it rains heavily during low tide, the sea reservoir outflow crest gates will be lowered to release the excess rainwater from the reservoir into the sea. If heavy rain falls during high tide, the outflow crest gates remain closed and large reservoir can be designed so as to absorb the floodwater shocks during high tide and release water out to sea only during low tides. The method further includes interlinking of these reservoirs at different locations across different rivers through underwater subsurface pipes for transferring water from one reservoir to another in shallow seawater (herein, we called this as Sarovar Mala; Sarovar means freshwater pond in Sanskrit and Mala means necklace around the land) – Thallak Gundurao Sitharam (2016).

10.4 Scenario in India

India is a peninsula surrounded by ocean on three sides. India enjoys a lot of monsoon rainfall because of its physical features. Indian peninsula receives an average 1172 mm of rainfall every year and this number has not changed much in the last 100 years. However, due to climate change, very intense rainfall events are causing floods in many regions in India. In some other areas, average annual rainfall is 300–650 mm, but it is very unreliable and the southwest monsoon accounts for most of the precipitation. Different regions receive different amounts of rainfall, varying from 200 mm to over 8000 mm. Despite that, we have a massive network of both perennial and nonperennial rivers that form a network within the country. India also has various water storing and conservation structures. India is the third largest dam building country in the world after the United States and China. As per the Central Water Commission of India, India has built 4964 large dams which are taller than 15 m since independence and majority of them (>80%) are earthen dams. Various storage structures lie at conflicting spots and hinder the socioeconomic condition of the people whenever there is a problem. Even these structures act as hotspots in both national and international borders. Interstate water disputes have only increased in the recent times. As per the Government of India information, there are eight major interstate water disputes in the country.

Due to various climatic factors, the total rainfall that should be received throughout the year is just received in 30–40 days. This would result in torrential rainfall and flash floods during monsoon. This precipitation of water ultimately would mix in the ocean without any use during the time of need. Close to 80% of the total precipitation received over Indian peninsula joins the ocean in just 3–4 months every year. Because of the lack of the storage structures within the country (even with 4877 large dams), we have a burning need of more storage structures to cater to water demands during times of drought. Although land-based reservoirs like dams have various advantages, they have their own disadvantages also. Land acquisition, clearing of forest area, loss of biodiversity, submergence of land and forests, diversion works during construction, etc., are some of them.

10.5 Benefits of coastal reservoir

There are many advantages of the sea-based reservoirs when compared with any conventional land-based reservoirs and they are as follows:

i. No harm to any of the river basins and no alteration to the river course (no temporary diversions as well)
ii. No disturbance to any forest cover and no submergence of land
iii. No physical displacement of people and their villages/towns
iv. Agriculture activity can be augmented
v. Coastal erosion can be minimized
vi. Groundwater recharge due to freshwater in estuarine areas
vii. Intrusion of saline water into wells will reduce
viii. Freshwater dredging will provide sand for construction
ix. Earthquake-resistant seawalls
x. Solar panels on the seawall – solar energy
xi. Tidal energy at the wall
xii. Roadways over the seawall, freshwater fishing, navigation and tourism
xiii. Real estate opportunities
xiv. Length and width of seawall – serve as a deepwater fishing harbour – benefit the fishing community.
xv. Increase in industrial, recreational and fishery activity around this freshwater

Table 10.1 gives the cost of water and cost of construction using coastal reservoirs in comparison to desalination, inland reservoirs and water recycling (from wastewater). Cost of water per 1000 L and cost for construction for storing 1 BCM of water have been estimated considering the costs of labour, materials and execution in India. Cost of water has been estimated at the location of coastal reservoir without considering the cost of pumping.

Table 10.1 Cost of water and cost of construction (1 US Dollar = Indian Rupees 70).

	Cost per kilolitre of water in Indian Rupees (Rs)	Cost of construction/billion cubic metres (in billion Indian Rupees)
Sea-based reservoirs	2–10 (sea level)	20
Inland reservoirs	30–100 (above the sea level)	100
Desalination	60–80 (sea level)	80

10.6 Coastal reservoirs in India

10.6.1 Thanneermukkom bund

Thanneermukkom bund (Fig. 10.2) was constructed in 1974 as part of Kuttanad low land development scheme and creation of freshwater reservoir in the coastal area of Kerala. Thanneermukkom saltwater barrier/bund is considered the largest mud regulator in the country and is in operation since 1976. It divides the Vembanad Lake into freshwater lake fed by the rivers draining into the lake and brackish water lake fed by ocean currents into low lands of Kuttanad. The four major rivers of Kerala, the Pamba, Meenachil, Achankovil and Manimala, flow into the region before they confluence into the Arabian Sea. It also has a major portion of the largest lake in Kerala, 'the Vembanad Lake'. Lake is fed by 10 rivers of which the above four major rivers form a main part. The regions receive good amount

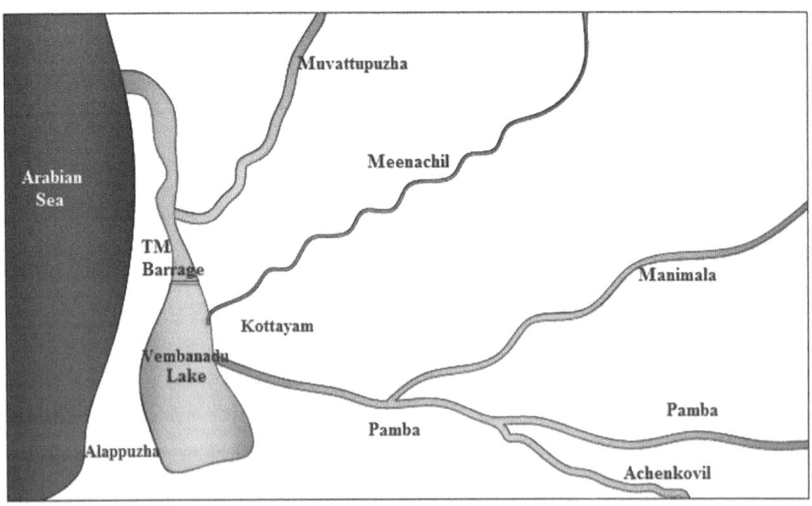

Figure 10.2 Schematic showing the location of Thanneermukkon Bund across Vembanad Lake and associated rivers.

of annual rainfall which is above 3000 mm and these four rivers bring large quantity of water into the lake before joining the sea. By constructing the saltwater barrier, a coastal reservoir having freshwater has been created for increasing agricultural activities in the area in addition to land development. However, there are reports of environmental and ecological damage such as rampant propagation of water hyacinth in freshwater and deterioration of brackish water fishing in the area. However, these are related to wrong operation of the reservoir and not fully functional plan of Thanneermukkom bund. The problems faced by fisherman and water hyacinth problems need to be addressed with innovative alternative schemes of operations. In fact this freshwater reservoir is need of the hour and needs to be restored to its capacity for supply of drinking water to nearby areas and also supply freshwater for irrigation in the low lands of Kuttanad, which helps farmers (Sitharam et al., 2018).

10.6.2 Kalpasar project

The state of Gujarat is listed as one of the high water-stressed states among various states in India. It has nearby 63 rivers, and 600 large and small dams are constructed in the state. The Kalpasar project is proposed at the Gulf of Khambhat, Gujarat, and addresses all the disadvantages as a regular inland reservoir has, and it also can solve the water crisis that is prevailing in the state. The Gulf of Khambhat Development Project is mainly water resources project involving creation of freshwater reservoir in the Gulf of Khambhat for meeting demand of irrigation and domestic and industrial water supply. Associated components related to the freshwater reservoir are use of top of the dam across the Gulf as a surface transport link, potential development of fisheries and reclamation of saline land around the freshwater reservoir.

The perspective view of the project is to create world's largest freshwater reservoir in sea, by constructing a 30 km long dam in the Gulf of Khambhat, it extends from north to south about 200 km and the width varies from 25 km at the inner end to 150 km at the outer mouth, covering an area of around 17,000 sq. km, of which only 2000 sq. km will be enclosed by constructing a dam across the Gulf between Bhavnagar and Dahej, which leads to storing more than 10,000 Million Cubic Meters (MCM) of surface water, i.e., 25% of volume of state's average annual rainwater inflow. The project aims to collect the runoff water from Sabarmati, Dhadar, Mahisagar, and Narmada rivers, which can be utilized for irrigation, industrial and drinking purpose. Fig. 10.3 shows the Kalpasar project location and detailed plan of the Gulf of Khambhat development. Further by using contour canals, water can be supplied to the entire Gujarat coast. Once the Gulf is shut, water levels inside the repository can be controlled while the tidal vacillation outside the supply proceeds and, henceforth, can be saddled for the age of tidal vitality. It is estimated that the transition of saltwater to target salinity of 0.1% will take 4 years, when the transformation from saltwater bay to freshwater reservoir starts. A 10-lane road link will also be set up over the dam, greatly reducing the distance between Saurashtra and South Gujarat by 225 km. A state government release said the Rs. 55,000 crore (US$ 11.7 billion) project to be completed by 2020, will have a vast freshwater reservoir with gross storage of 16,791 million cubic metres of water, 64 km long dam across the Gulf of Khambhat connecting Ghogha in Bhavnagar with Hansot in Bharuch District. State government of Gujarat, India, has signed a memorandum of understanding with Korean consortium to build the first ambitious seawall project in the Gulf of Khambhat in the first quarter of 2016 (Kalpasar, 2016). Once constructed it will be one of the large freshwater reservoirs in the sea with highest priority for irrigation and drinking water for Saurashtra and Central Gujarat regions of India.

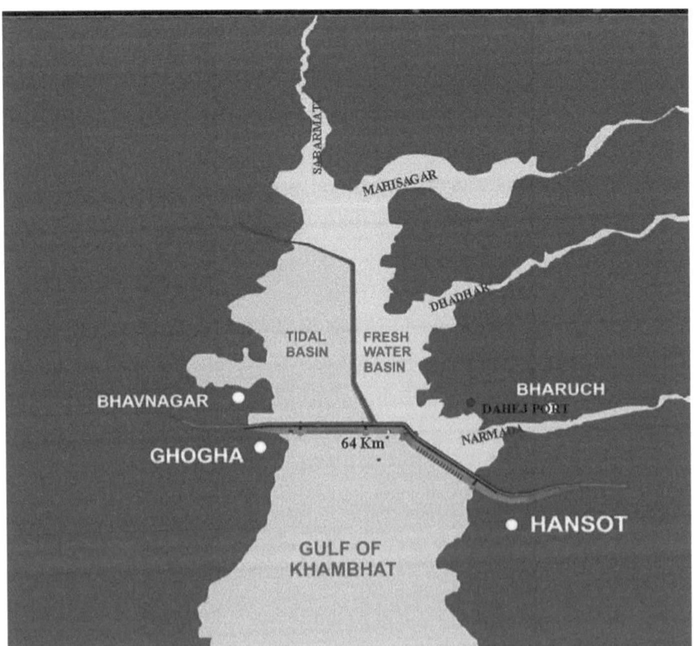

Figure 10.3 Kalpasar – proposed 64 km dam across the Gulf of Khambhat, Gujarat.

10.6.3 Feasibility study of coastal reservoir at Mangalore from floodwaters of Netravati River

Authors have carried out a detailed feasibility study with large multidisciplinary team on developing a sustainable water source for Banglore and Mangalore using the coastal reservoir as a storage area for Netravati/Gurupura river floodwaters, Karnataka, India. The principal objective of the project was to assess the feasibility of storing freshwater in a reservoir along the coast, by building a sea dike in the Arabian Sea. On one side, the dyke will ensure the required quantity of freshwater flow from the Netravati to the reservoir, without being drained completely into the sea. On the other side, the sea dike will prevent the seawater from entering the reservoir avoiding salt contamination of the freshwater supply (Fig. 10.4). The study addresses the coastal urban watershed management of Mangalore city and agricultural land along the river and the coast, in the context of the proposed project of building a sea dike for the creation of a freshwater reservoir impounding river floodwaters (BWSSB report, 2017; Kolathayar et al., 2017; Parthasarathy et al., 2019).

Figure 10.4 Proposed scheme for the coastal reservoir project in Mangalore.

It is imperative that a small percentage of runoff of Netravati is more than sufficient to cater the water requirements of Mangalore and Bangalore. From the data of last few decades, there are hardly any chances of scarcity of water availability in Netravati River. The concept of coastal reservoir emerges as the best solution to meet the future water demands of both Bangalore and Mangalore. The average annual runoff in Netravati is 388.5 TMC ft. As per Bangalore Water Supply and Sewerage Board (BWSSB), shortfall in demand in Bangalore for 2051 is 26.16 TMC ft, which is 7% of average annual runoff through Netravati. In 2021, the shortfall in demand is only 8 TMC ft, which is just 2.1% of the runoff. There is hardly any chance of silting of reservoir as the average annual sediment load in Netravati River is 0.04 TMC ft. The water quality parameters in Netravati and Gurupura Rivers are in tolerable limits for safe drinking water and the water can thus be directly utilized without any major treatment. Benefiting sectors due to coastal reservoir are agriculture, prawn culture and sand dredging. The possible renewable energy generation implies that the system will be self-sufficient in energy to meet the energy demands for pumping, lighting and other requirements. Consultation and inputs from renewable energy developers would strongly support in the planning for cost-effective and sustainable technology installations for renewable energy generation. The coastal reservoir will not affect the built-up area. The land use indicates that built-up area is less, while the vegetation cover is more. Also, due to freshwater reservoir, a reduction in salinity of water is expected after the construction of coastal reservoir, and its effects on the vegetation must be preassessed and analyzed.

10.6.4 Sarovar Mala (Garland of Reservoirs for India)

Sarovar Mala (Fig. 10.5) is a concept of linking sea-based/coastal reservoirs (Sarovar's in the ocean) at the mouth of rivers (where river joins the ocean in peninsular India). The main objective is to create a large body of freshwater for developmental activities including drinking/agricultural activities in the coastal area and increase water storage per capita in the coastal areas. This is a strategic plan to increase the water storage per capita in India along the coast by storing the excess floodwaters from major rivers in southern Indian Peninsula. These Sarovars can be connected by pipelines in shallow ocean waters along the coast (by using a shallow water spud barge which has side way lowering).

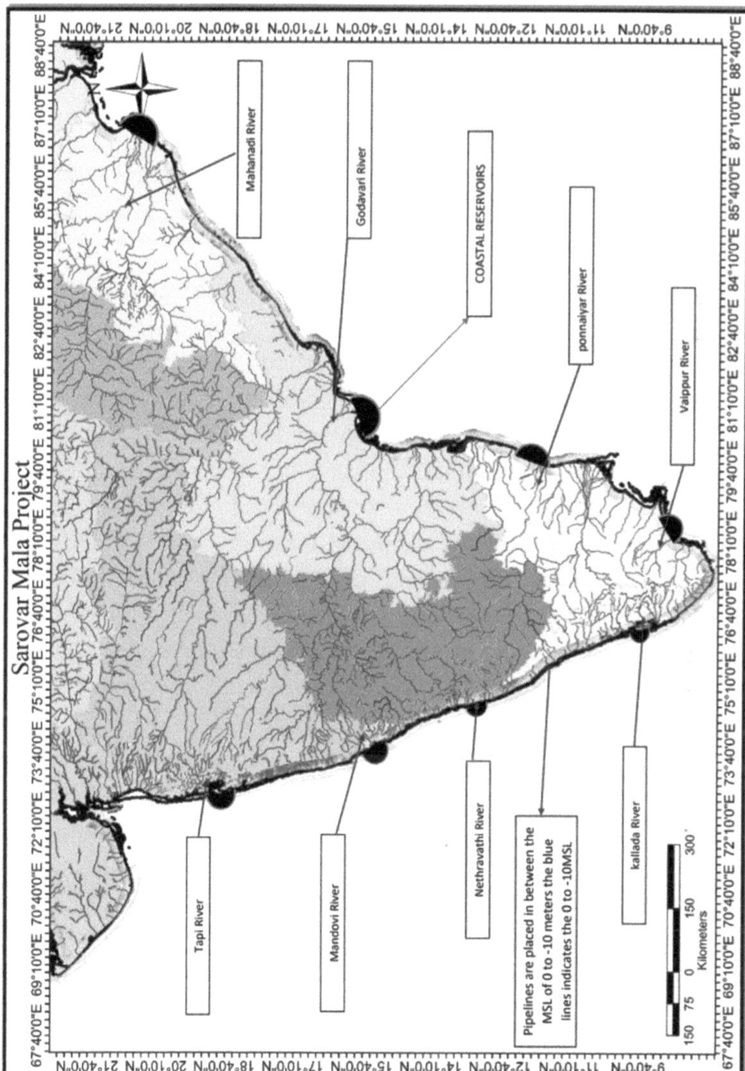

Figure 10.5 Sarovar Mala, a chain of freshwater reservoirs along the coastal line of India.

Along with the initiative of the Government of India to modernize India's Ports, this series of freshwater reservoirs (Sarovars) will aid the coastal area development along with new ports. The increased freshwater availability in the coastlines can be developed to contribute to India's growth. Sarovar Mala provides a good connectivity for nonperennial rivers and will provide a storage plan for freshwater from the excess river floodwaters and one can develop the industrial clusters and hinterland up to a radius of 100 km resulting in Sarovars becoming the drivers of economic activity in coastal areas.

10.7 Concluding remarks

India receives most of its freshwater during a 4-month monsoon period during which time the lakes are filled, rivers experience floods and groundwater is recharged. But due to onset of climate change, the rainfall pattern changed, resulting in extreme variability of the rainfall experienced. While some rivers receive huge quantities of rain resulting in extreme flooding, other rivers and waterbodies are bone dry. Indian population over the decades has increased so much that the density of population and dwindling forest coverage is not allowing creation of fresh land-based reservoirs. The original Indian proposal of interlinking of rivers to overcome the ill effects of variable rainfall is practically dead and impossible to implement under the changed circumstances and huge cost of the project. Considering all the above, the solution lies in storing floodwater in sea-based reservoirs as this does not submerge lands which are of immense value. If heavy rain falls during high tide, the outflow crest gates remain closed and large reservoir of about >100 TMC ft (2.83 BCM) can be designed so as to absorb the floodwater shocks during high tide and release water out to sea only during low tides. Even the sand, silts and salts can join the ocean partly through the sea-based reservoir. Sea-based reservoir can be constructed in shallow waters at appropriate locations close to the mouth of the river along with a barrage at one or two ends. A detailed study is needed even when considering saltwater intrusion through river estuarine, bays and gulfs. Sarovar Mala by connecting coastal reservoirs around the Indian peninsula will be a sustainable solution for Indian water requirements specifically in the coastal areas. More species of fish can find these ponds home. Coastal reservoirs have the potential to increase India's freshwater supply for generations to come by efficiently taping the excess floodwaters from rivers flowing into Arabian Sea and Bay of Bengal.

References

BWSSB Report, 2017. Coastal Reservoir at Mangaluru to Impound Netravati River Flood Waters – A Feasibility Study in India. Submitted to BWSSB, Govt of Karnataka, p. 178.

Kalpasar, 2016. The Gulf of Kambhat Development Project. https://kalpasar.gujarat.gov.in/mainpage.htm.

Kolathayar, S., Sitharam, T.G., Ramkrishnan, R., Rao, R.S., 2017. Feasibility of creating a fresh water reservoir in the Arabian Sea impounding the flood waters of Netravathi River. Advances in Sustainable Urbanization 2 (2), 38–42.

Kolathayar, S., Sitharam, T.G., Yang, S., 2019. Coastal reservoir strategy to enhance India's freshwater storage by impounding river flood waters: a detailed overview. Water Supply 19 (3), 703–717.

Parthasarathy, C.R., Sitharam, T.G., Kolathayar, S., 2019. Geotechnical considerations for the concept of coastal reservoir at Mangaluru to impound the flood waters of Netravati River. Marine Georesources and Geotechnology 37 (2), 236–244.

Sitharam, T.G., Kolathayar, S., Yang, S., Krishnan, A., 2018. Concept of a geotechnical solution to address the issues of sea water intrusion in Ashtamudi Lake, Kerala. In: Civil Infrastructures Confronting Severe Weathers and Climate Changes Conference. Springer, Cham, pp. 238–246.

Thallak Gundurao Sitharam, 2016. Smart Water Network to Store River Flood Water in Sea Based Reservoirs and a Method of Interlinking These Reservoirssss (Sarovar Mala). India patent no. TEMP/E-1/44294/2016-CHE.

Yang, S.-Q., 2010. Coastal Reservoirs Can Harness Stormwater. Water Engineering, Australia, pp. 25–27.

Yang, S.-Q., 2015. Coastal Reservoir- A technology that may dominate future water supply. Journal of Water Resource and Hydraulic Engineering 4 (4), 388–397.

Yang, S.-Q., Kelly, S., 2015. The use of coastal reservoirs and SPP strategy to provide sufficient high quality water to coastal communities. Journal of Geoscience and Environment Protection 3, 80–92. https://doi.org/10.4236/gep.2015.35010.

Yang, S.-Q., et al., 2005. Coastal Reservoirs. Singapore Patent No. 200504653-7.

Yang, S., Liu, J., Lin, P., Jiang, C., 2013. Coastal reservoir strategy and its applications. In: Wurbs, R. (Ed.), Water Resources Planning, Development and Management. Intech, United States, pp. 95–115.

CHAPTER 11

An alternative method to solve water scarcity in Adelaide – apply a coastal reservoir strategy in the Lower Lakes

Jianli Liu[1], Muttucumaru Sivakumar[2], Shu-Qing Yang[2], Brian G. Jones[3]
[1]Centre for Coastal Reservoir Research, University of Wollongong, Wollongong, NSW, Australia; [2]School of Civil, Mining and Environmental Engineering, University of Wollongong, Wollongong, NSW, Australia; [3]School of Earth, Atmospheric and Life Sciences, University of Wollongong, Wollongong, NSW, Australia

11.1 Introduction

The Murray–Darling river basin (MDRB) covers 1 million km^2, which is about 1/7 of the Australian land area. The Murray and Darling rivers merge in New South Wales and drain to the sea via Lake Alexandrina. Their runoff comes from New South Wales, Queensland, Victoria and the Australian Capital Territory where the average annual rainfall is about 530 mm/year over the basin.

A severe drought occurred in MDRB in the 1930s. Adelaide could not use the river water for irrigation and domestic purposes because of seawater intrusion into the Lower Lakes in dry years. In 1930–40, five barrages were constructed at the outlet of Lake Alexandrina. In some sense, this was the first attempt in Australia to construct a coastal reservoir because the design purpose of the barrages was to stop seawater intrusion and to maintain a high freshwater level in the lake, which can be considered as a 'first-generation' coastal reservoir. But because of the inappropriate design and size of the coastal reservoir, these purposes were only partially achieved. This paper investigated the feasibility of building a suitable size 'second-generation' coastal reservoir in Lake Alexandrina to contain and clarify water by using gates operations.

11.2 Water supply and demand analysis for Adelaide

The MDRB has an area of 1,042,730 km^2 and includes parts of Queensland, New South Wales, Victoria, South Australia and the Australian Capital Territory (Fig. 11.1A) (CSIRO, 2008). Lake Alexandrina Fig. 11.1B is a

Sustainable Water Resource Development Using Coastal Reservoirs
ISBN 978-0-12-818002-0
https://doi.org/10.1016/B978-0-12-818002-0.00011-3

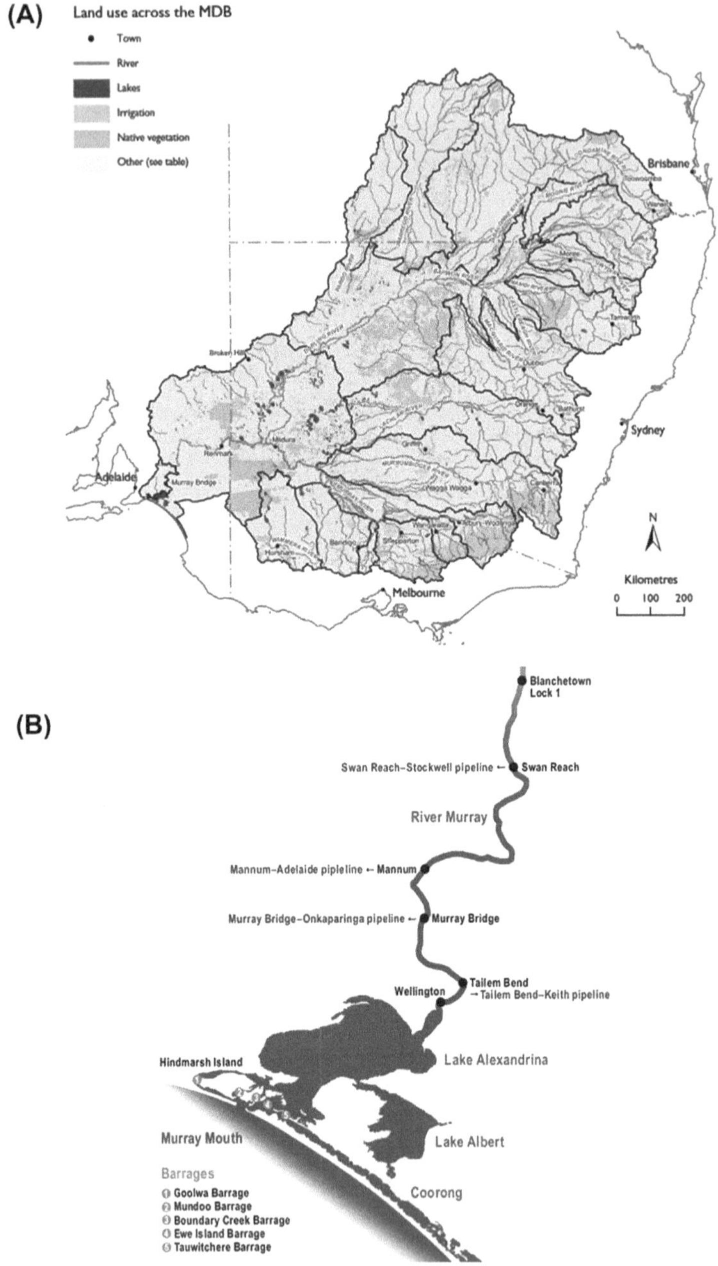

Figure 11.1 (A) Map of Murray–Darling Basin regions and (B) the Lower Lakes in South Australia (MDBA, 2011). *((A) From CSIRO (2008). Water availability in the Murray-Darling Basin. A report to the Australian Government from the CSIRO Murray-Darling Basin Sustainable Yields Project.CSIRO, Australia. 67pp. © Copyright CSIRO Australia).*

broad and shallow (mean depth 2.86 m, maximum depth 4.75 m), well-mixed, regulated water body, with a surface area of approximately 650 km^2 and a volume of approximately 1620 GL at +0.7−0.75 m AHD (Australia Height Datum). Five barrages separate Lake Alexandrina from the Coorong lagoon and the Murray estuary, which are the Goolwa, Mundoo, Boundary Creek, Ewe Island and Tauwitchere barrages (Gell and Haynes, 2005). Lake Albert is a terminal lake of the Murray River linked to Lake Alexandrina by a narrow channel (the Narrung Narrows) between Point Malcolm and Narrung Peninsula, through which it receives most of its inflows.

Fig. 11.2 shows the water needs for Greater Adelaide in the future according to the Government of South Australia. There were two conditions which included dry year conditions and forecast demand. In 2050, about 430 GL of water will be needed for Greater Adelaide in dry year conditions. For forecast demand, 360 GL will be needed for Greater Adelaide in average years. Thus, 360−430 GL will be used as the basis for the coastal reservoir design.

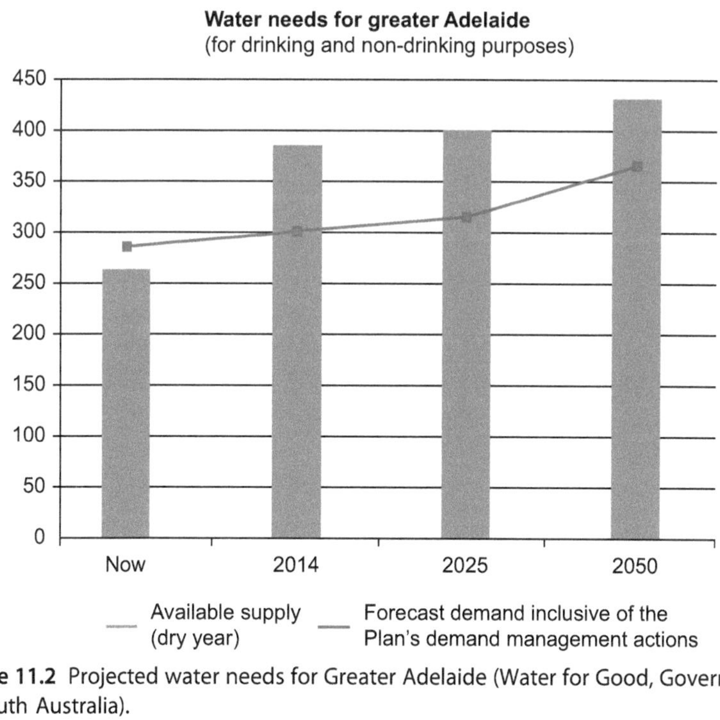

Figure 11.2 Projected water needs for Greater Adelaide (Water for Good, Government of South Australia).

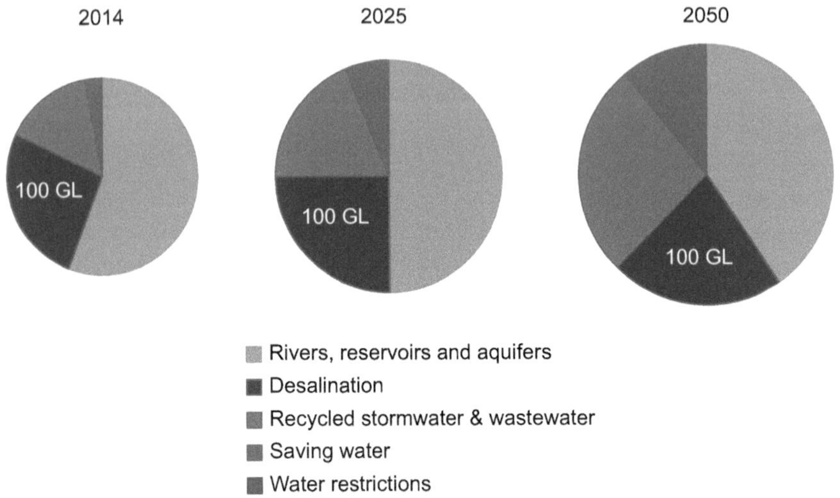

Figure 11.3 Water sources for Greater Adelaide in the future (Government of South Australia, 2010).

Water source distributions for 2014, 2025 and 2050 are shown in Fig. 11.3. In 2050, the volume of water sourced from the Murray River, local reservoirs and aquifers, as well as the desalinated water is assumed to remain the same. The main differences were the increases from recycled stormwater and wastewater and saving water.

As indicated above, the possible amount of water needed in 2050 will be 360–430 GL/year. In Greater Adelaide, the target for 2050 is to achieve the capacity to recycle at least 60 GL/year of stormwater for nondrinking purposes, and a minimum of 75 GL/year of the wastewater generated in South Australian urbanized areas will be recycled for nondrinking purposes. Saving water by restricted use is projected to be around 50 GL/year. So the water demand from a coastal reservoir can be based on these numbers (see Table 11.1). As stated above, a total of 360–430 GL/year of water will be needed to ensure a sustainable population water consumption rate. Therefore, estimated demand in 2050 for an average rainfall year is 360 GL/year, whereas the mean forecast demand for dry year conditions is 430 GL/year.

As the cost of desalination is very high, a coastal reservoir strategy is suggested to augment desalination. For the 'water provision with a coastal reservoir (GL/year) – average rainfall year' and 'water provision with a coastal reservoir (GL/year) – dry year', there was no need to save water to guarantee household water consumption (Table 11.1). By 2050, approximately 78–148 GL/year of additional water will be needed for Adelaide

Table 11.1 Measured and predicted water usage distribution in 2008, 2014 and 2050.

Method	Water in 2008 (GL/year)	Water provision in 2014 (GL/year)	Water provision in 2050 by plan (GL/year)	Water provision with a coastal reservoir (GL/year) – average year	Water provision with coastal reservoir (GL/year) – dry year
Water supply from rivers, reservoirs and aquifers	194	147	147	147	147
Recycled stormwater and wastewater	30	45	75 + 60	75 + 60	75 + 60
Water restrictions	32	0	0	0	0
Saving water	10	8	50	0	0
Desalination	0	100	100	0	0
Water demand	0	0	0	78	148
Total	266	300	432	360	430

through other sustainable water sources. Therefore, the coastal reservoir should be able to store 78–148 GL/year. In this study, it was taken as 150 GL/year for the coastal reservoir's water supply for Adelaide.

Based on these data, a 'second-generation' coastal reservoir has been designed in the lake as shown in Fig. 11.4 where the reservoir area is 150 km^2, the depth range of the coastal reservoir is 3.5–4 m and storage is about 520–580 GL.

11.3 Applying coastal reservoir strategy in Lake Alexandrina

11.3.1 Coastal reservoir gates and mesh

This study used MIKE software modelling suite by DHI (2014). Four reservoir gates are set up in the Lower Lakes for good operation of the coastal reservoir. The upstream flow would be through Gates 1 and 2 to the Lower

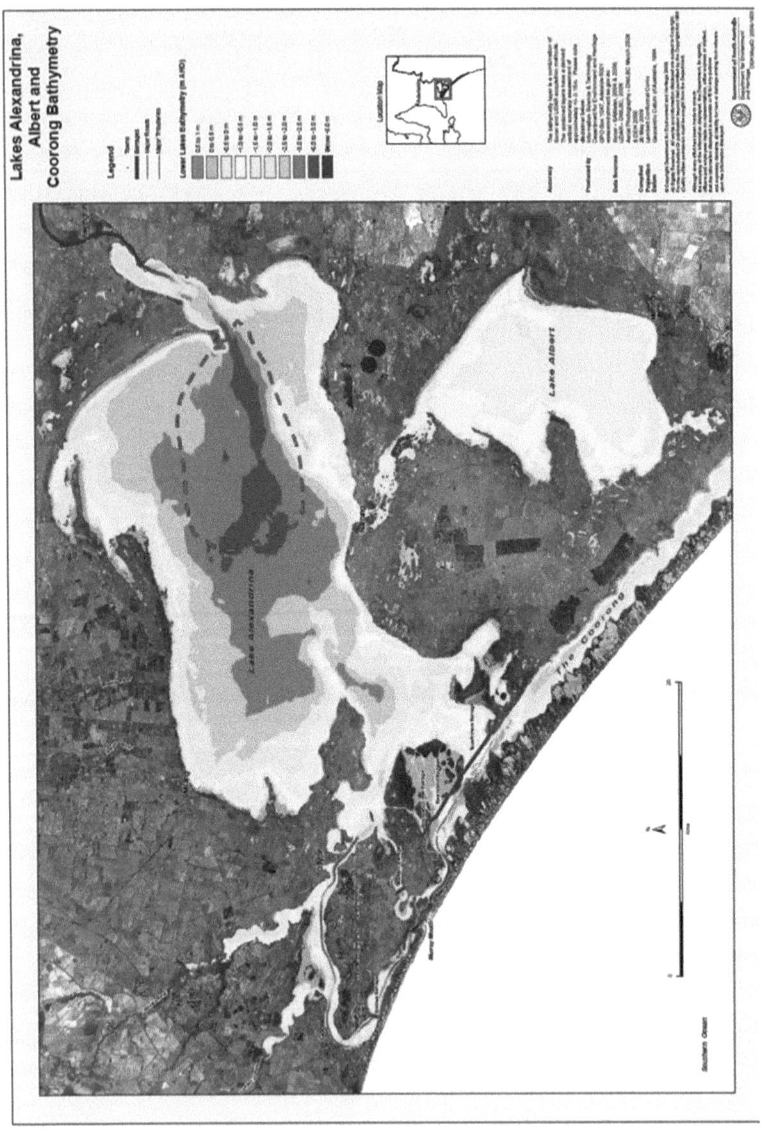

Figure 11.4 Proposed coastal reservoir enclosing the deep water by a dike (*red line* [*grey line* in print version]).

Lakes (outside the coastal reservoir). Murray River water from upstream would pass through Gate 3 into the coastal reservoir, and Gate 4 is built for water to flow from the coastal reservoir to the Lower Lakes (Fig. 11.5).

11.3.2 Applying the coastal reservoir strategy in Lake Alexandrina

Heneker (2010) researched the characteristics of historical inflows to Lake Alexandrina and barrage outflows. Based on Lake Alexandrina inflow statistics, the 10th percentile annual lake inflow is 1110 GL (based on 1891/92–2007/08 historical records) and 920 GL (based on 1975/76–2007/08 records). The mean value for annual lake inflow is 5780 GL (based on 1891/92–2007/08) and 4960 GL (based on 1975/76–2007/08 records). Analyzing the flow data for the Murray River (MDBA, 2013) shows that the annual inflow of 1122 GL occurred in 2002 is close to the 10th percentile annual lake inflow, whereas in 1998 with an annual inflow of 5036 GL it is nearest to the 50th percentile annual lake inflow. Also, 2007 was chosen to verify extreme drought conditions. These three typical year flows (2007 – drought, 2002 – 10th percentile and 1998 – 50th percentile) are applied to investigate the coastal reservoir strategy.

11.3.2.1 Extreme drought condition (2007)

Models were set up and compared the simulation results before and after building a coastal reservoir using 2007 conditions.

- Gate operation

For the coastal reservoir model simulation for year 2007 inflows, the gates' operating conditions are shown in Fig. 11.6, in which 1 stands for gate open, 0 stands for gate closed. Gates 1 and 2 would be closed from January to August, while the Murray River water has a salinity below 0.3 PSU (Fig. 11.7), and Gate 3 would be open to let the low salinity water into the coastal reservoir. Thus, the coastal reservoir admits only good quality water. Gate 4 would be opened or closed depending on whether the coastal reservoir has filled up. From September to December, the salinity of water from the Murray River increases up to 0.4 PSU. At this time, Gate 3 would be closed, and Gates 1 and 2 would be opened to let Murray River water flow directly into Lake Alexandrina.

- Comparison between before and after the coastal reservoir (2007 conditions)

Six sites were chosen for comparison of modelling results and are shown in Fig. 11.8. Points 1, 2, 4 and 5 represent the southeast, northeast,

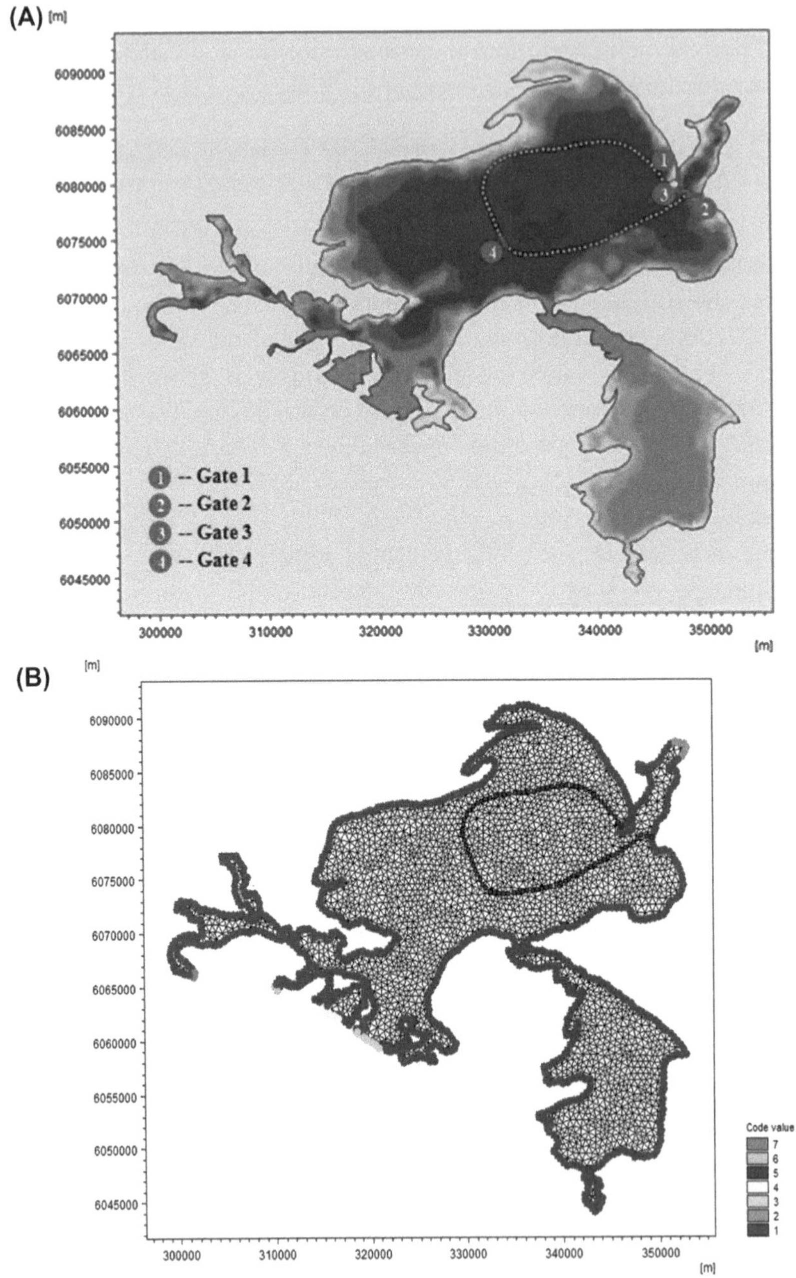

Figure 11.5 Schematic diagram for the coastal reservoir showing (A) gate locations; and (B) the computational grid.

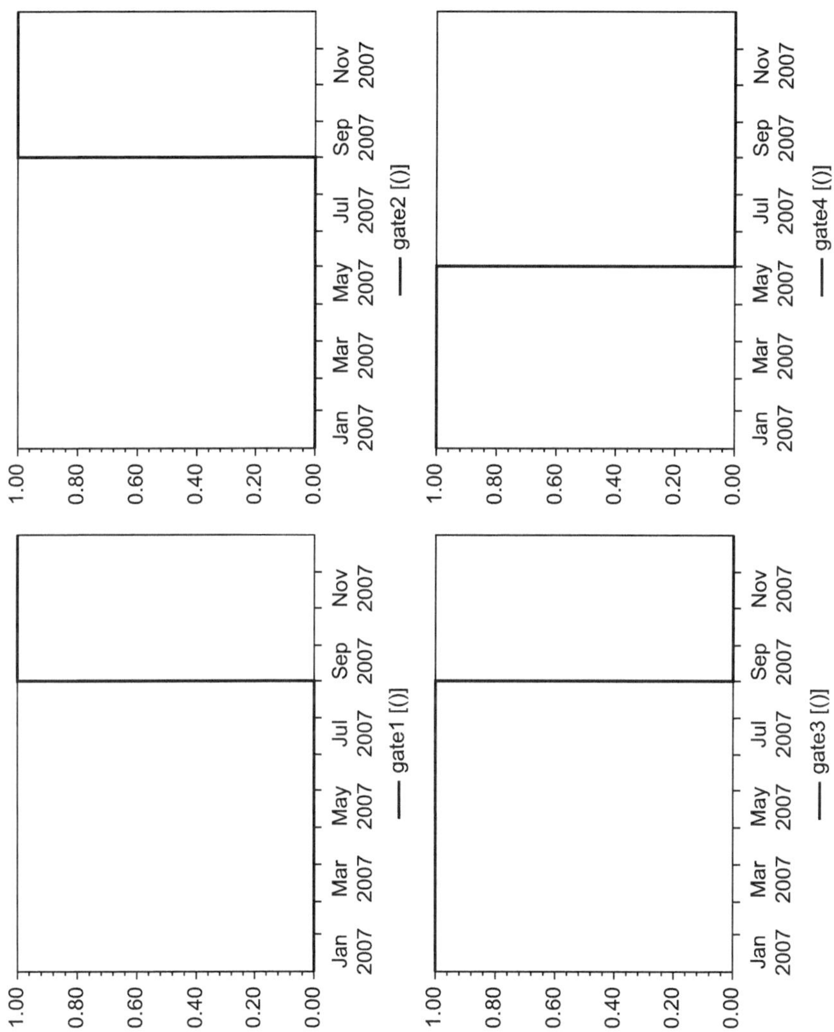

Figure 11.6 Gates operations for the coastal reservoir using 2007 conditions.

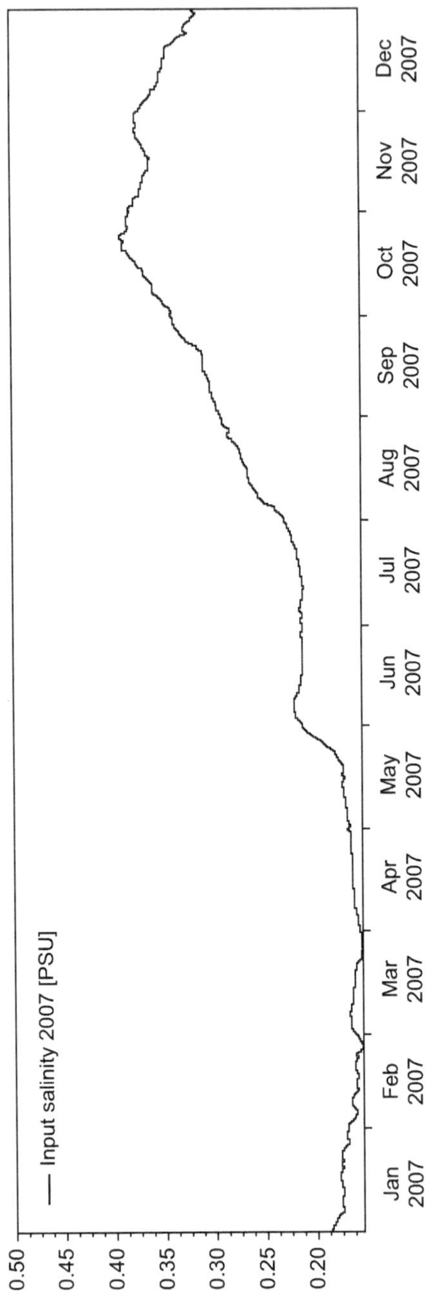

Figure 11.7 Lock 1 salinity variation in 2007.

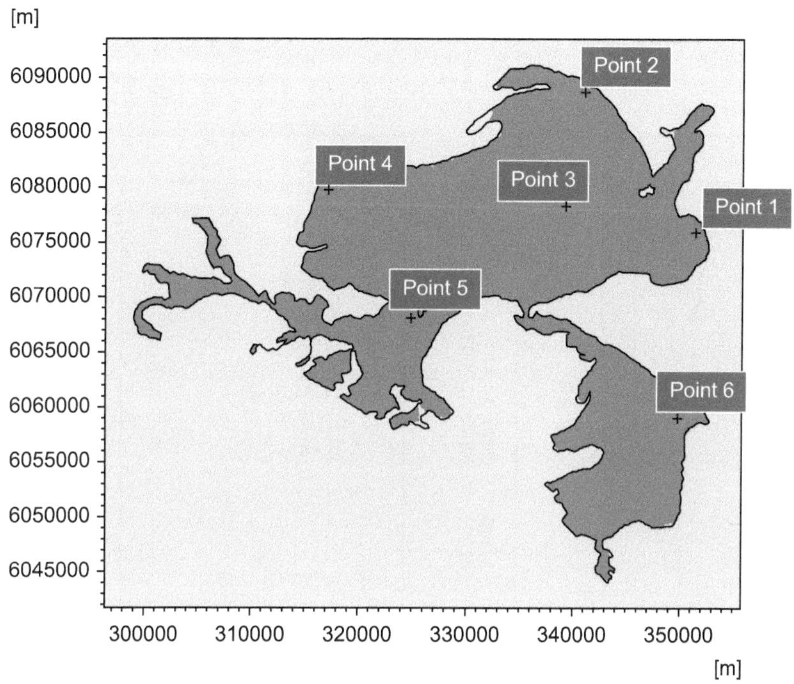

Figure 11.8 Site locations for model results comparison.

northwest and southwest parts of Lake Alexandrina. Point 3 is in the middle of Lake Alexandrina within the proposed coastal reservoir. Point 6 is a site in Lake Albert. Also, these six sites in the Lower Lakes have monitored data, which make it easy to get historic hydrological and salinity data and will be convenient for monitoring in the future. The salinity comparison results are as shown in Fig. 11.9.

From the salinity comparison (Fig. 11.9), it is seen that salinity variation follows a similar trend before and after building the coastal reservoir except at Point 3. The difference in salinity at Point 1 after building the coastal reservoir is marginally higher than that before building the coastal reservoir from January to August. From September to December, salinity at Point 1 after building the coastal reservoir is lower than that before building the coastal reservoir. This is because Gates 1 and 2 are closed from January to August while the Murray River water flows into the coastal reservoir; from September to December, Gates 1 and 2 are open, and Gate 3 is closed, which means that more Murray River flows into the northern and southern parts of Lake Alexandrina (except the part of coastal reservoir) than before building the coastal reservoir. At Point 2, the maximum increase is only 0.1 PSU. Salinity at Point 3 after building the coastal reservoir is obviously

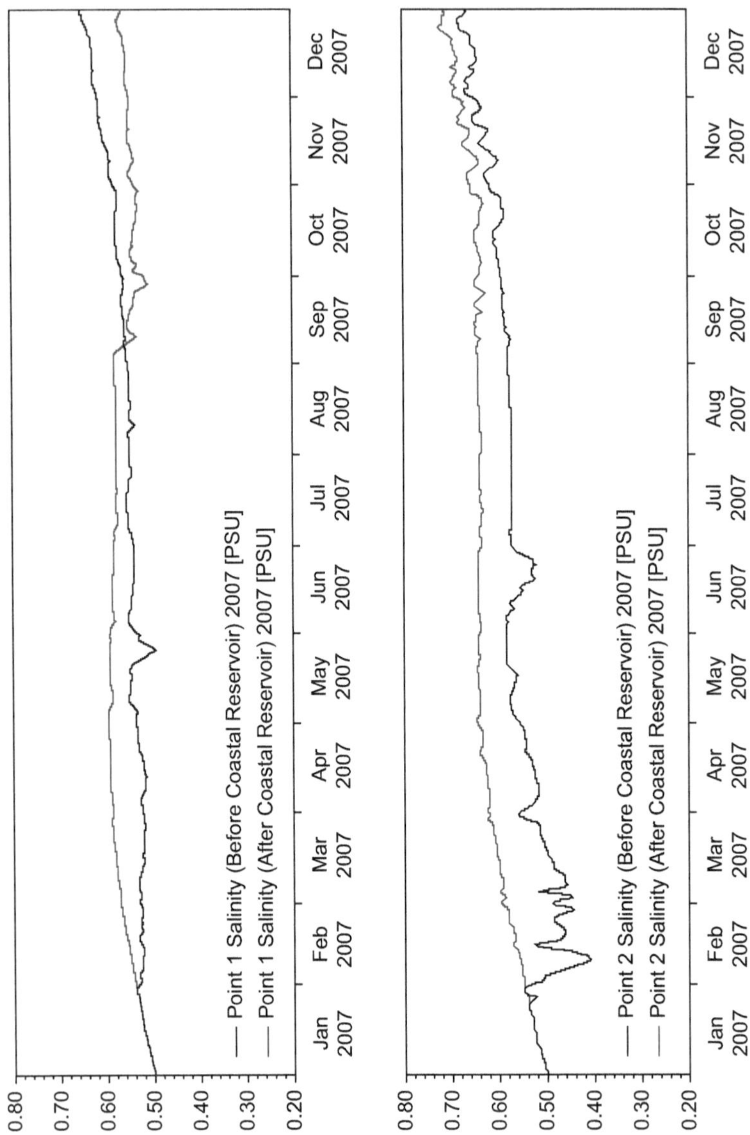

Figure 11.9 Salinity comparison between before and after the coastal reservoir based on 2007 conditions.

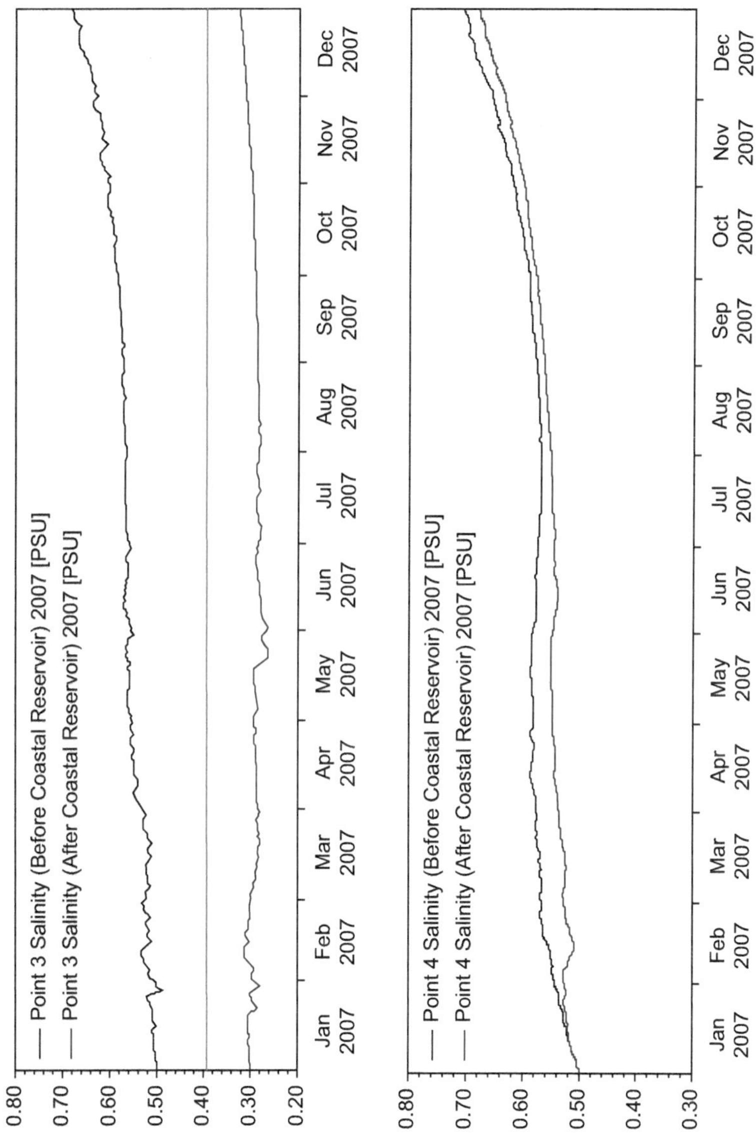

Figure 11.9 Cont'd

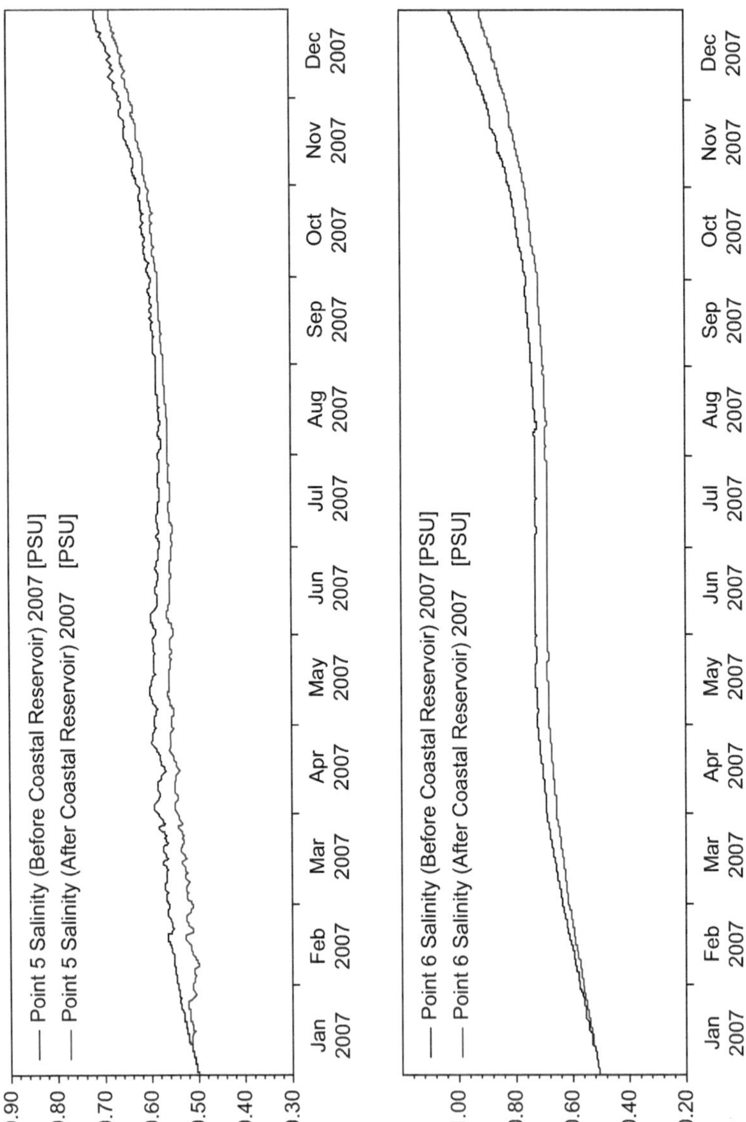

Figure 11.9 Cont'd

lower than that before building the coastal reservoir, which is maintained at around 0.3 PSU compared with 0.5 PSU before building the coastal reservoir. This is because the salinity in the coastal reservoir is lowered through the gate operations that let more low salinity Murray River water enter the coastal reservoir. The potable drinking water standard in South Australia for salinity is 0.4 PSU. Salinity at Point 3 is below 0.4 PSU at all times, which is readily acceptable for potable water supply. Salinity at Points 4, 5 and 6 after building the coastal reservoir is also found to be lower than that before building the coastal reservoir, which means that building the coastal reservoir has a good effect on the Lower Lakes.

11.3.2.2 10th percentile condition using 2002 data

Models were run to simulate the situations before and after building the coastal reservoir in the Lower Lakes for 2002 data.

- Gate Operation

The rules for the four gates are shown in Fig. 11.10. From January to April, a large volume of Murray River flows into the Lower Lakes. Salinity at this period is at around 0.3 PSU (Fig. 11.11). Gates 1 and 2 would be

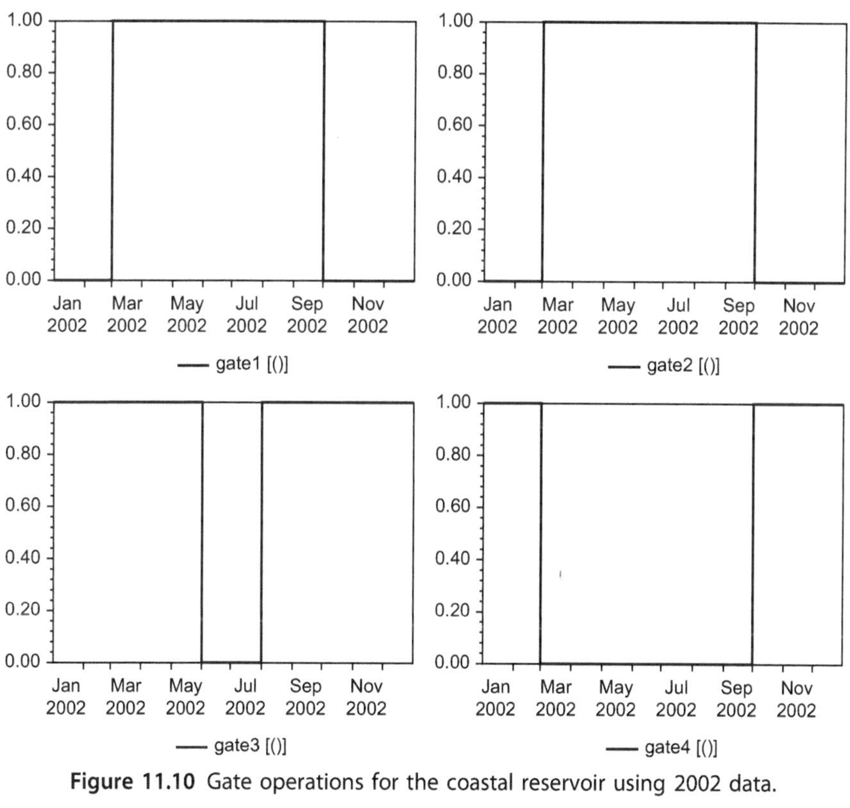

Figure 11.10 Gate operations for the coastal reservoir using 2002 data.

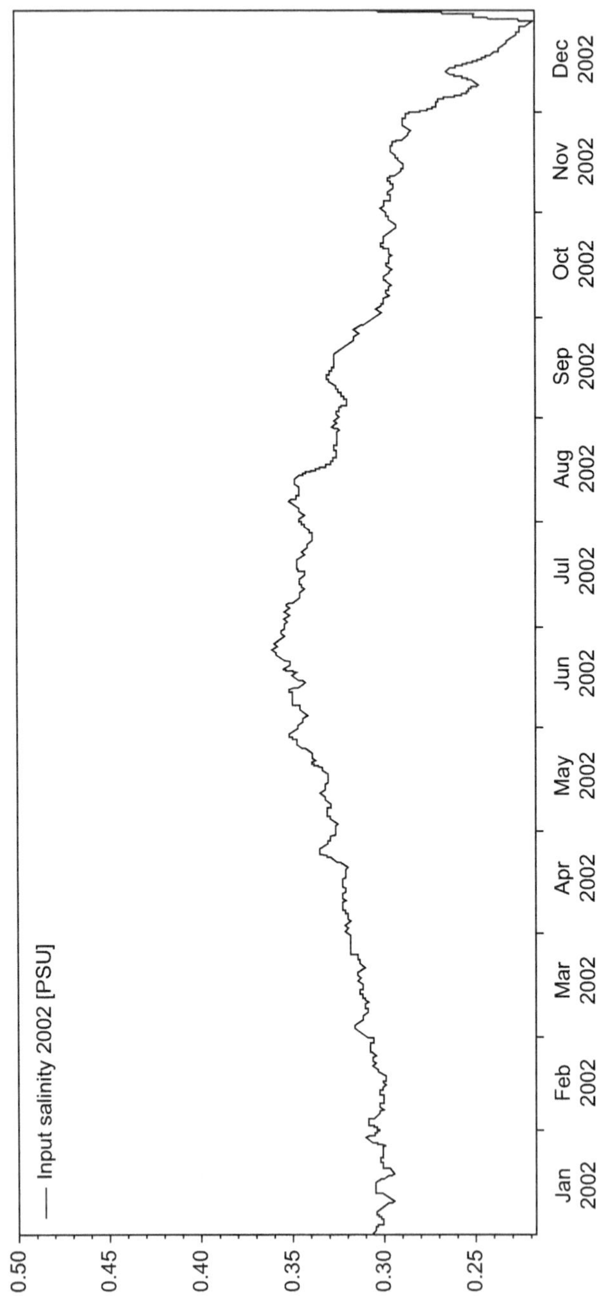

Figure 11.11 Lock 1 salinity variation in 2002.

closed, and Gates 3 and 4 would be open in January and February to let freshwater enter and flush the coastal reservoir. Then from March to September, Gates 1 and 2 would be opened to let Murray River flow into Lake Alexandrina. Gate 3 would be opened from January to March, when there is a large volume of lower salinity inflow. Gate 4 would be closed from March to September to retain the Murray River flow. In June and July, there is less volume of water flowing into the Lower Lakes (about 18 m^3/s), and the salinity is up to 0.35 PSU. Gate 3 would be closed to let this water into Lake Alexandrina. From August, the water flow from the Murray River is increasing, and the salinity is also lower than that in June and July. Gate 3 can be opened to replenish water storage in the coastal reservoir. From October to December, the inflow salinity decreases from 0.28 PSU to 0.17 PSU (Fig. 11.11), so Gate 3 would be opened to let the lower salinity water enter and flush the coastal reservoir (Fig. 11.12).

- Comparison between before and after the Coastal Reservoir (2002 conditions)

The same sites are chosen for comparing results (Fig. 11.8). A comparison of salinity at the six chosen sites before and after building the coastal reservoir is shown in Fig. 11.12.

From Fig. 11.12, it is seen that nearly all salinity values at six different sites after building the coastal reservoir are lower than those before the coastal reservoir. For Point 1, salinity after building the coastal reservoir is slightly higher than before building the coastal reservoir during January and February, mainly because of Gates 1 and 2 being closed, and there is no water flowing into the Lower Lakes. Yet for the period from March to October, Gates 1 and 2 would be open, and Gate 3 would also open except in May and June. In the latter 2 months, higher volume of water would flow into Lake Alexandrina, which causes the salinity at Point 1 after building the coastal reservoir to be lower than before building the coastal reservoir. For Point 3, which is in the coastal reservoir, the salinity can be maintained at around 0.3 PSU. For other points in the Lower Lakes, salinity after building the coastal reservoir is lower than before building the coastal reservoir during most of the year. For Points 4 and 5, salinity is much lower in February. This is because Gate 4 would be opened at that time. When Gate 3 is opened to let lower salinity, Murray River flows into Lake Alexandrina, and Gate 4 would also be opened to let the good quality water flush the coastal reservoir, which can decrease the salinity of the area around Gate 4 and improve the environment. Salinity at Point 6 after building the coastal reservoir is also lower than before building the coastal reservoir. One

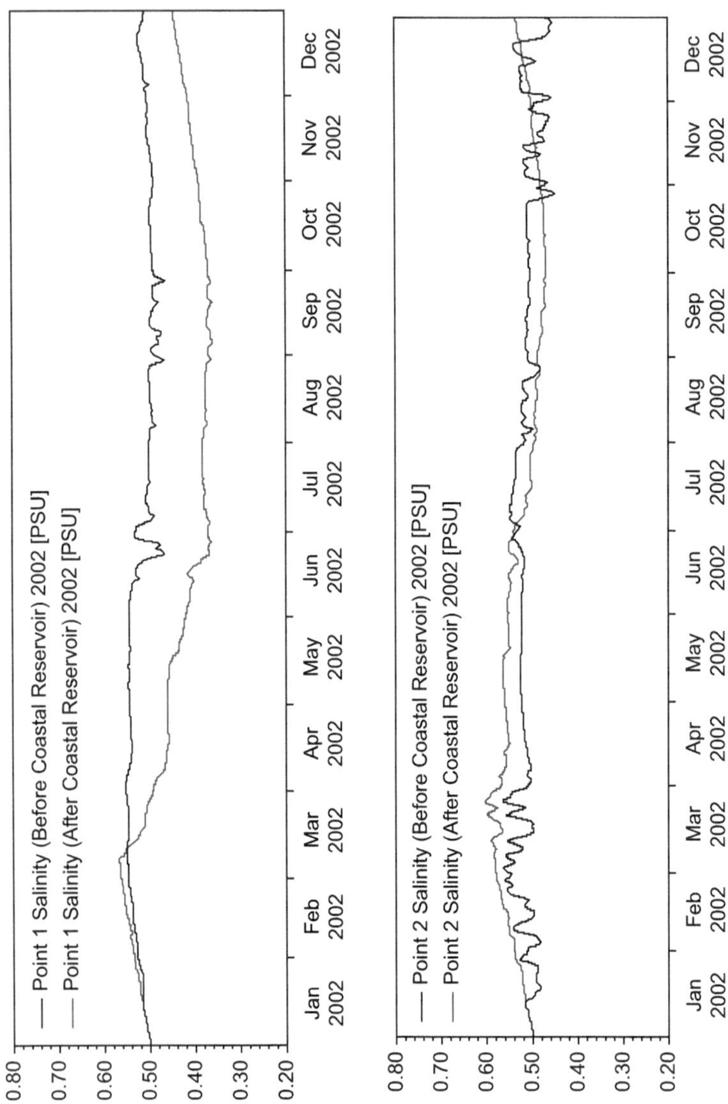

Figure 11.12 Salinity comparison between before and after the coastal reservoir based on 2002 conditions.

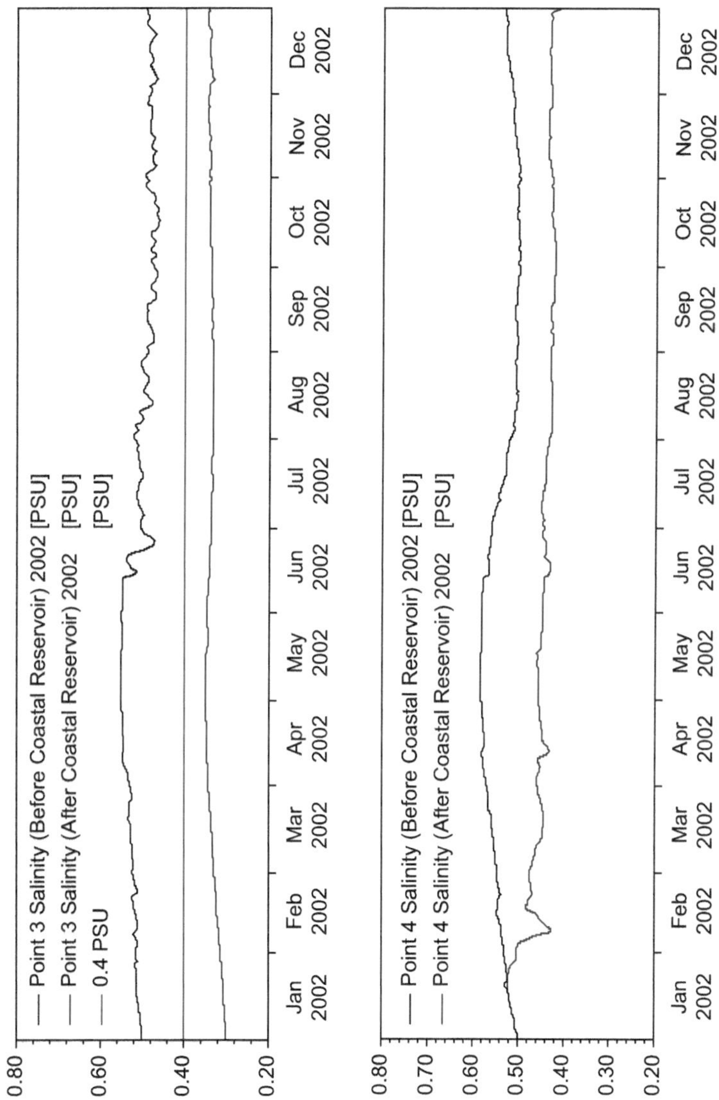

Figure 11.12 Cont'd

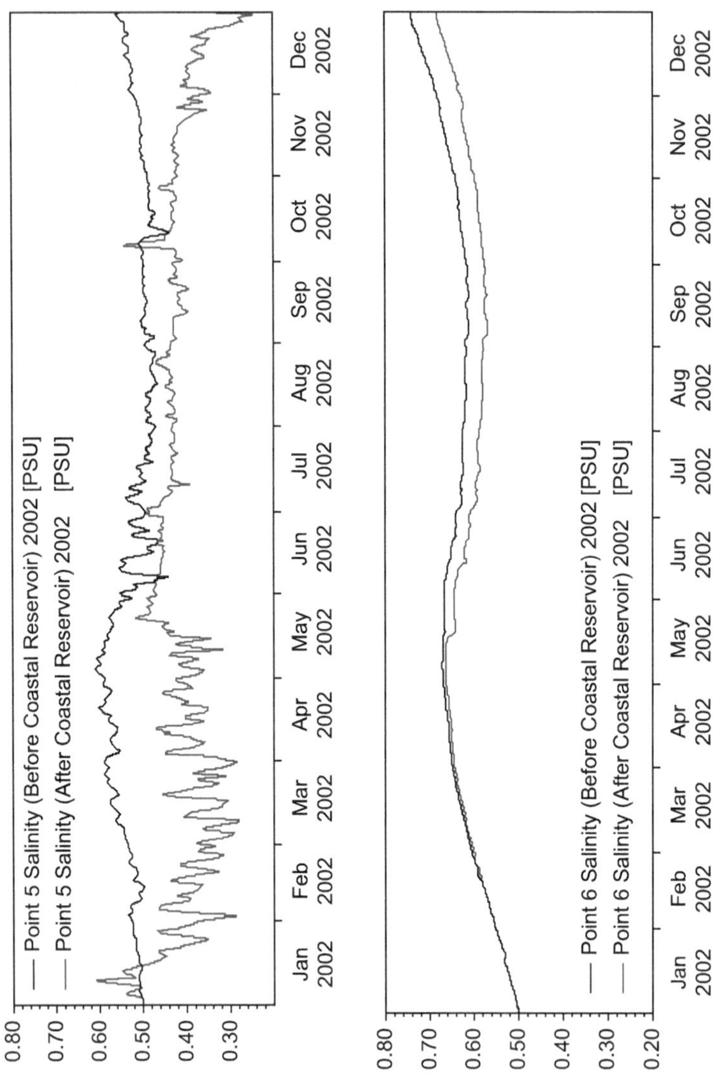

Figure 11.12 Cont'd

reason for the salinity to decrease is that the salinity in the coastal reservoir would be around 0.3 PSU, and the gates' rational operation would store more good quality water in the Lower Lakes.

11.3.2.3 50th percentile condition using 1998 data

- Gate operation

The rules for the four gates are shown in Fig. 11.13.

Fig. 11.13 shows the operational patterns of the gates. The operation for Gates 1 and 2 is the same. Gate 3 would be closed while Gates 1 and 2 are open in January because the input salinity is above 0.3 PSU (Fig. 11.14). Higher salinity water would flow through Gates 1 and 2 into Lake Alexandrina and is blocked from entering the coastal reservoir. From February to May, inflow salinity is under 0.3 PSU. Gates 3 and 4 would be opened to let the freshwater flush the coastal reservoir. As the inflow volume is large, Gates 1 and 2 are also opened during these months, which import good quality water into the Lower Lakes and prevent overflowing of the coastal

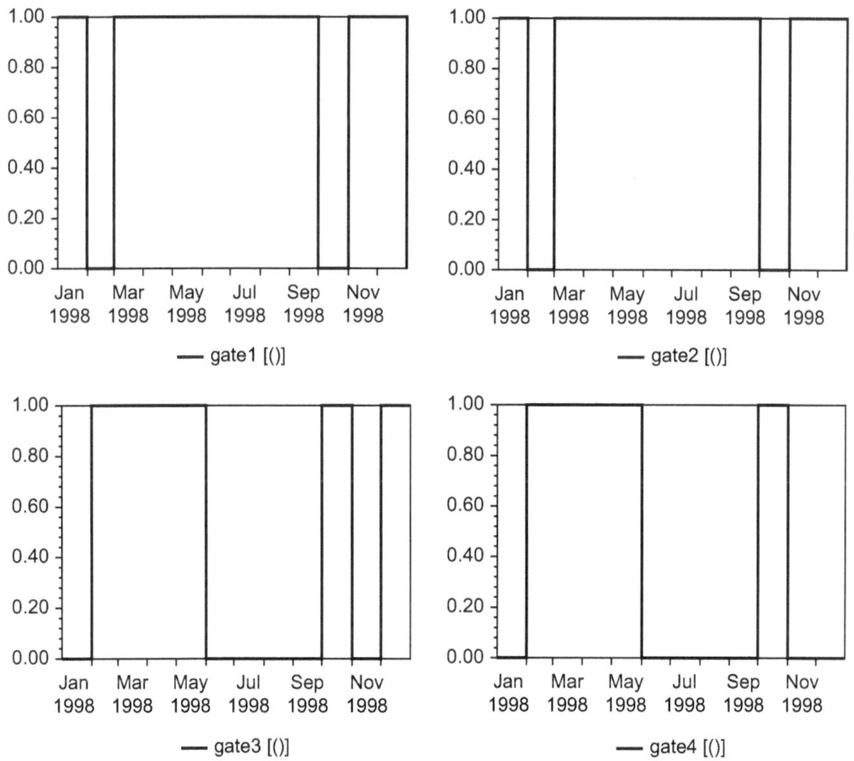

Figure 11.13 Gates opened and closed for the coastal reservoir using 1998 data.

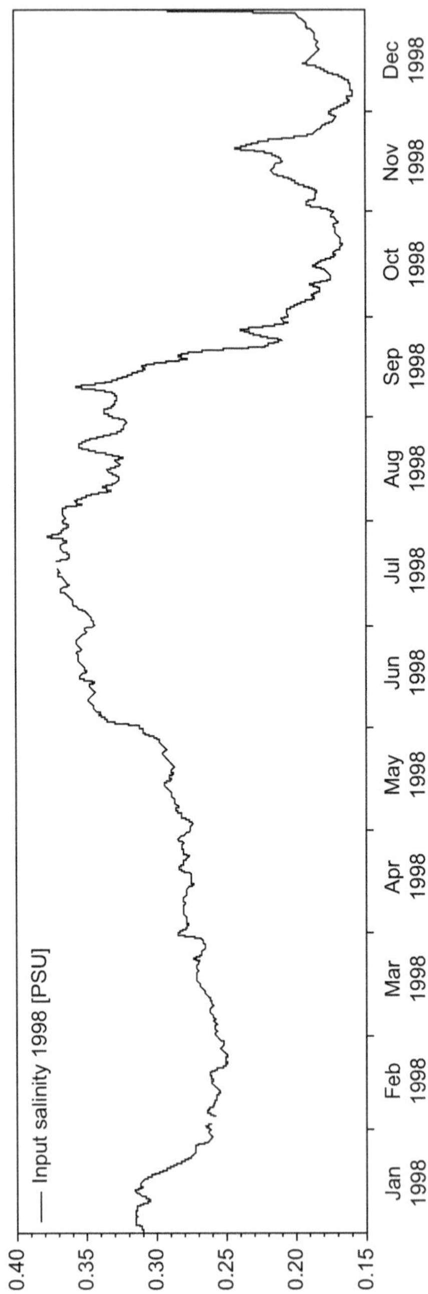

Figure 11.14 Lock 1 salinity variation in 1998.

reservoir. From June to September, inflow salinity is higher (around 0.35 PSU). Gates 3 and 4 would be closed, and Gates 1 and 2 would be opened to let Murray River water flow into the Lower Lakes. In October, inflow salinity has decreased to 0.17 PSU. Gate 3 would be opened, and Gates 1, 2 and 4 would be closed to let lower salinity water enter and be stored in the coastal reservoir. In November, inflow salinity is getting higher, and the Murray River flow volume is large; the coastal reservoir would have already stored enough water in October. At this time, Gates 3 and 4 would be closed to let Murray River flow into Lake Alexandrina through Gates 1 and 2. For December, inflow salinity is lower again (approximately 0.2 PSU), and also the Murray River flow volume is large (200 m^3/s). Gates 1, 2 and 3 would be opened at the same time to let Murray River flow into Lake Alexandrina and to the coastal reservoir. Gate 4 would be closed to keep more freshwater in the coastal reservoir.

- Comparison between before and after the Coastal Reservoir (1998 conditions)

A comparison of salinity at the six chosen sites before and after building the coastal reservoir is shown in Fig. 11.15.

For conditions like 1998, Gates 1 and 2 are open for most of the year, and with an annual inflow of 5000 GL, it is easy to flush and store good quality water in the coastal reservoir. The days required for opening Gate 3 are less than for an annual inflow of around 1000 GL (the year 2002). When Gate 3 is closed, the whole upstream water flows into Lake Alexandrina, which improves the lake's environment. For Point 1 (Fig. 11.15), the salinity after building the coastal reservoir is lower from January to the middle of March compared with salinity before building the coastal reservoir. The salinity would be marginally higher from April to June and then became lower from July to December. Gates 1 and 2 would be closed in February, which would cause the salinity to go up. The same principle also applies to Point 2 in northeastern Lake Alexandrina (Fig. 11.15). Salinity at Points 1 and 2 would be a little bit higher than before the coastal reservoir when Gates 1 and 2 are closed, and Gate 3 is open. But as the flow volume is large, Gates 1 and 2 would not be closed for a long time. As soon as Gates 1 and 2 are reopened, the salinity in the southeast part (Point 1) and northeast part (Point 2) would become lower again. Fig. 11.15 shows that the salinity in the coastal reservoir (Point 3) can be kept at around 0.3 PSU. At Point 4, the salinity is unaffected because Point 4 is located at the western edge of Lake Alexandrina and it is not affected too much by a 5000 GL annual flow (1998). For Point 5, its salinity after building the coastal reservoir is much lower than that before

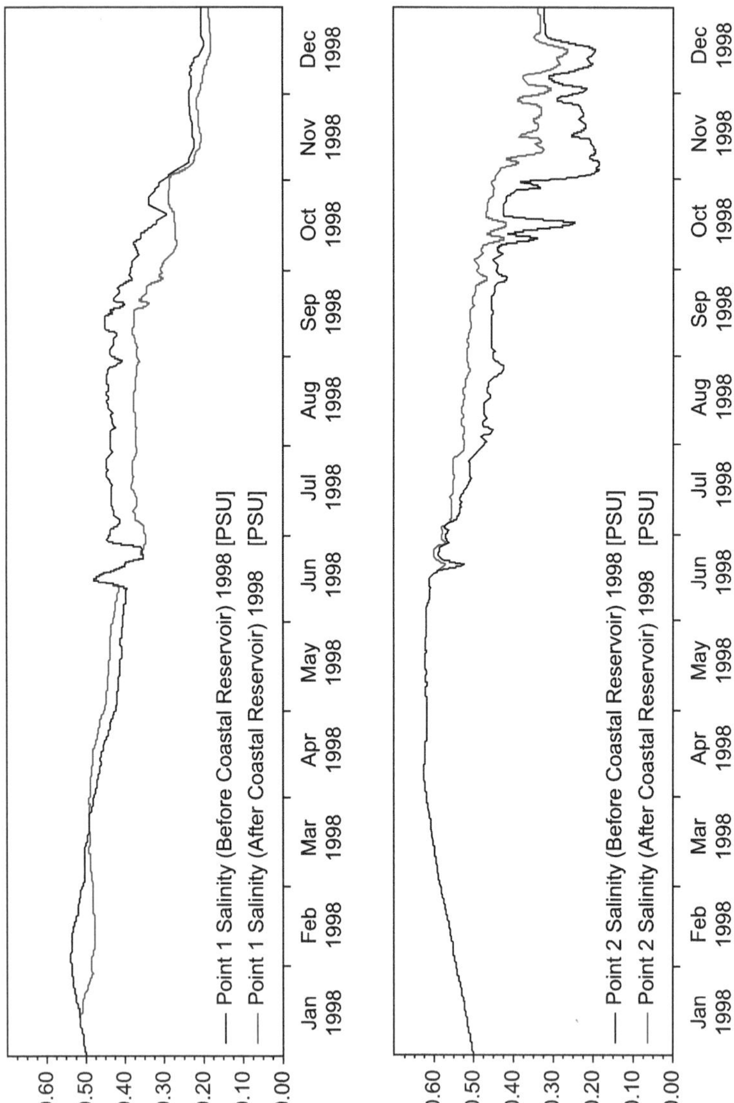

Figure 11.15 Salinity comparison between before and after the coastal reservoir based on 1998 conditions.

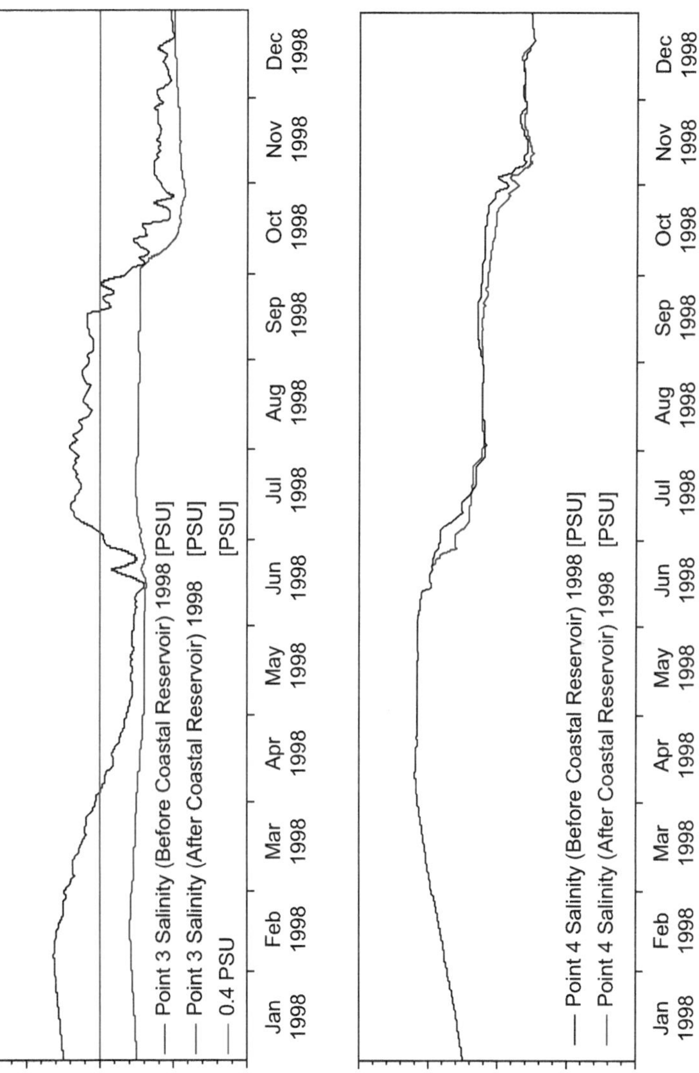

Figure 11.15 Cont'd

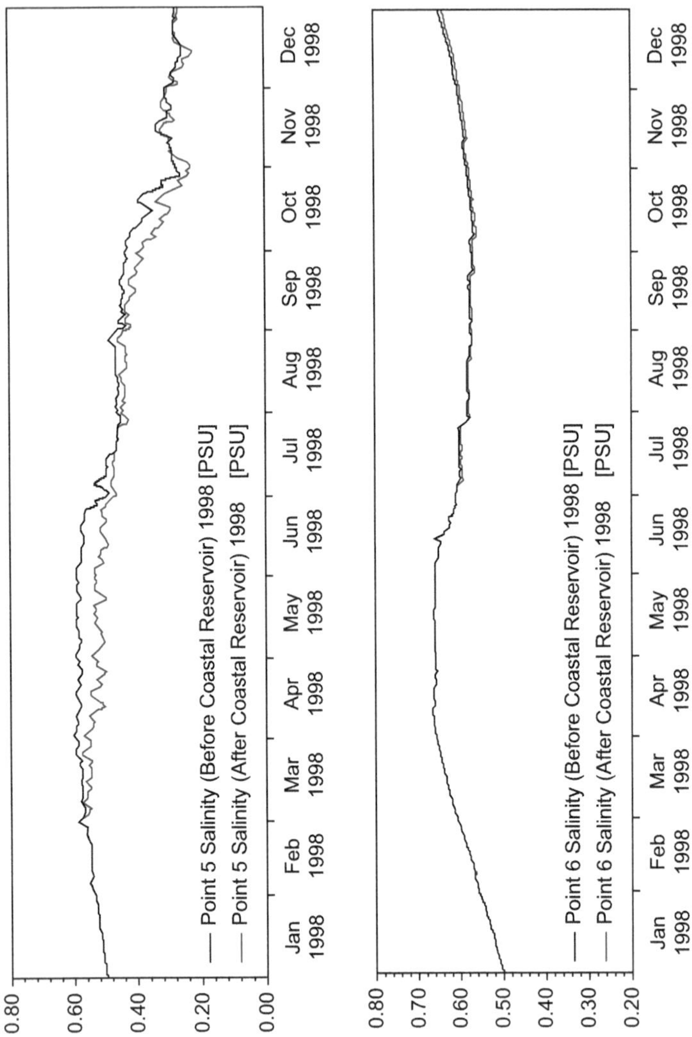

Figure 11.15 Cont'd

coastal reservoir. This is because when Gate 4 is open, there is a good freshwater flow from Gate 4 to Point 5. For Point 6, salinity after building the coastal reservoir does not show much change compared to that before the coastal reservoir. So for the situation like 1998 (around 5000 GL/year), salinity after building the coastal reservoir is basically the same or better than before building the coastal reservoir.

11.3.2.4 Velocities comparison with and without coastal reservoir (using 2007 as an example)

Using 2007 as an example, Fig. 11.16 shows the velocity comparisons (m/s) for the six sites (Fig. 11.8) between before and after building a coastal reservoir, where *Vp'* means velocity after building coastal reservoir and *Vp* means velocity before building a coastal reservoir. If $Vp' - Vp > 0$, it means velocity at a given site after building coastal reservoir is faster than velocity before building a coastal reservoir.

For Point 1, from late January to March, velocities after building coastal reservoir are slower than those before building a coastal reservoir. For other months, velocities after building coastal reservoir are faster than those before building a coastal reservoir. So for most months of the year, velocities after building coastal reservoir are faster than those before building a coastal reservoir, which has a positive effect on this part of the lake environment. For Point 2, at most times of the year, velocities after building coastal reservoir are faster than those before building the coastal reservoir, which has a positive effect on this part of the lake environment. For Point 3, at most times of the year, velocities after building the coastal reservoir are slower than those before building a coastal reservoir, which is as expected because the coastal reservoir is a relatively enclosed space. For Point 4, velocities after building the coastal reservoir are sometimes faster and sometimes slower than those before building the coastal reservoir. For Point 5, at most times of the year, velocities after building the coastal reservoir are faster than those before building the coastal reservoir, which has a positive effect on this part of the lake environment. For Point 6, velocities after building the coastal reservoir are smaller than those before building the coastal reservoir. The differences in velocities generally range from ± 4 to 8 mm/s.

Overall, for the northwest and northeast parts of Lake Alexandrina, velocities after building the coastal reservoir are faster than those before building the coastal reservoir. For other parts of Lake Alexandrina, velocities after building the coastal reservoir and before building the coastal reservoir did not show much difference. For Lake Albert, velocities after building the coastal reservoir are a bit slower than those before building the coastal reservoir.

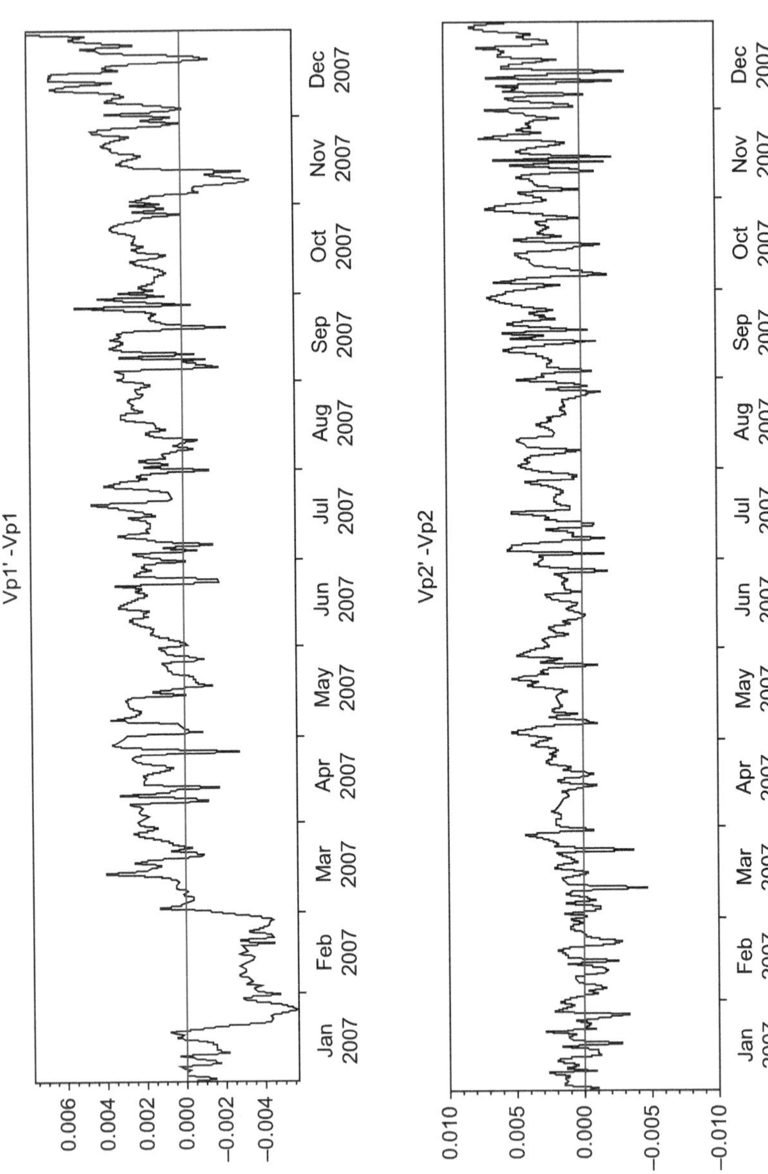

Figure 11.16 Velocity comparisons before and after building coastal reservoir based on 2007 data.

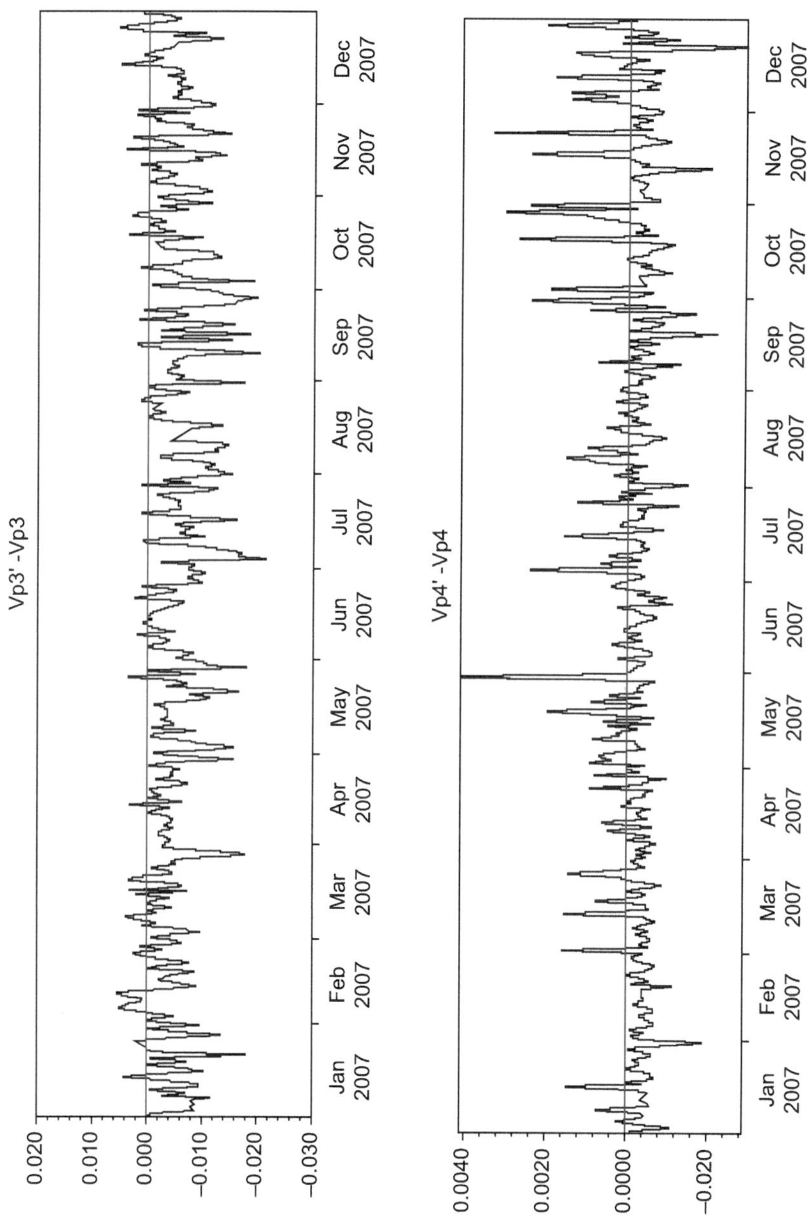

Figure 11.16 Cont'd

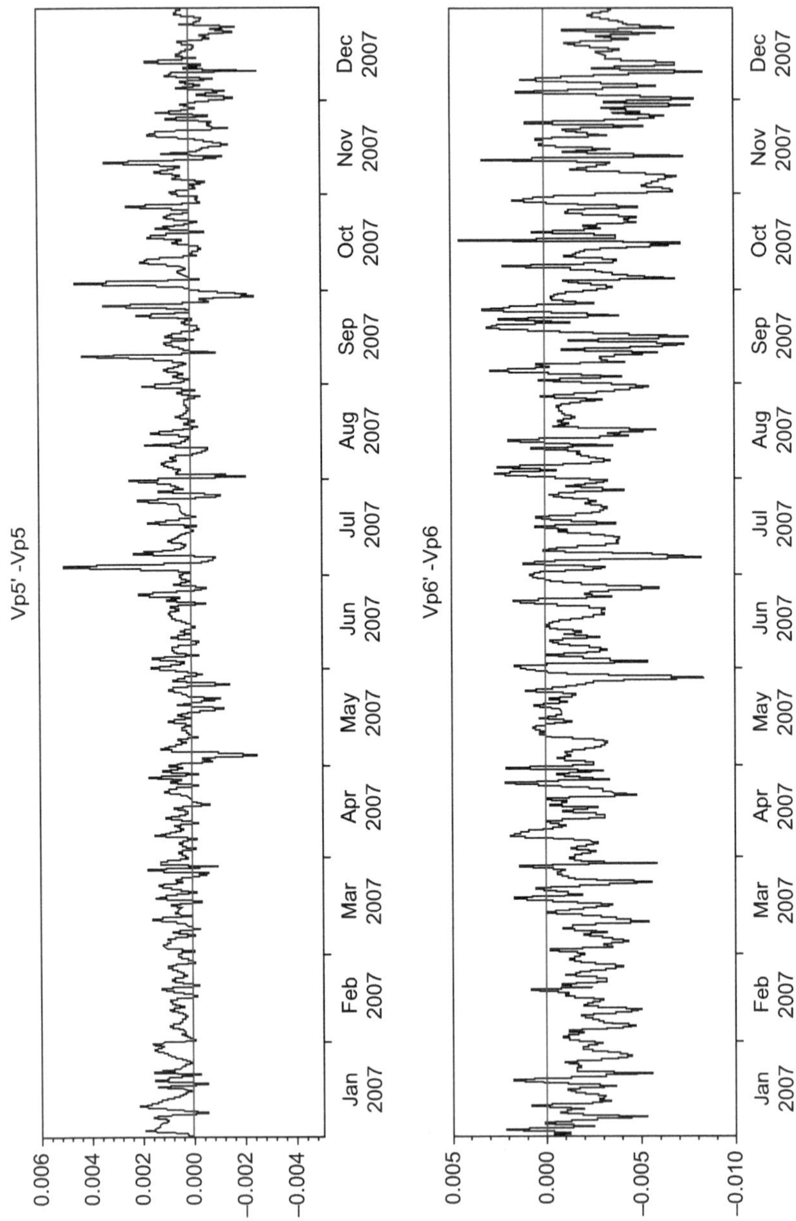

Figure 11.16 Cont'd

11.3.2.5 Salinity and outflow comparisons with and without a coastal reservoir

The average salinity and outflow with and without a coastal reservoir for the three selected years are summarized and given in Table 11.2.

For the year 2007, as all the five barrages were closed, there was no flow coming out through the five barrages. For the condition without a coastal reservoir, the average salinity for Lake Alexandrina was around 0.552 PSU. For the condition with the coastal reservoir, the average salinity for the lake is calculated to be 0.497 PSU, so the average salinity with the coastal reservoir is lower than that without the coastal reservoir.

For the year 2002, the outflow for the condition without coastal reservoir was 185 GL as recorded (Heneker, 2010) while the barrage outflow for the condition with the coastal reservoir would be 35 GL as 150 GL would be pumped and provided to Adelaide. For the 2002 conditions, the coastal reservoir would cause a drop in flow into the Coorong. The average salinity in 2002 before the coastal reservoir was around 0.503 PSU, while the average salinity after the coastal reservoir would be around 0.497 PSU, which is lower than that before the coastal reservoir.

The outflow from barrages in 1998 was recorded as 3315 GL (Heneker, 2010). As the coastal reservoir provides 150 GL/year for Adelaide, after the coastal reservoir, the outflow for the barrages would be 3165 GL. This is 4.5% of the 3315 GL, which is a small proportion, so under this condition, pumping 150 GL would not make any significant changes to flows into the Coorong.

Table 11.2 is a rough summary. For practical conditions, the coastal reservoir would already contain enough water (at 0.7 m AHD water level) in high flow years, so there is more water stored in the Lower Lakes. The barrage outflow would, therefore, be much greater than 35 GL, which would reduce the effect of the modelled decreased outflow on the Coorong.

Table 11.2 Comparison with and without coastal reservoir.

Year	Item	Without coastal reservoir	With coastal reservoir
2007	Average salinity (PSU)	0.552	0.497
	Outflow (GL)	0	0
2002	Average salinity (PSU)	0.503	0.462
	Outflow (GL)	185	35
1998	Average salinity (PSU)	0.501	0.435
	Outflow (GL)	3315	3165

11.4 Conclusions

The Lower Lakes of the MDRB is at the terminus of the Murray River and is about 100 km southeast of Adelaide, South Australia. This study set up a numerical model by using MIKE software for the Lower Lakes and applied a 'second-generation' coastal reservoir strategy in the Lower Lakes to solve future water shortages for Adelaide. Hydraulic, hydrodynamic and salinity processes in the lake were modelled to simulate the impact of a coastal reservoir. The results show that if a second-generation coastal reservoir with appropriate gate operations to admit only good quality water is built within Lake Alexandrina, it could provide an additional 150 GL/year of freshwater for Adelaide. The salinity within the coastal reservoir can be kept at around 0.3 PSU. The construction of such a second-generation coastal reservoir, as a new strategy, is potentially beneficial in many parts of Australia as well as in overseas countries.

References

CSIRO, 2008. Water Availability in the Murray: A Report to the Australian Government from the CSIRO Murray-Darling Basin Sustainable Yields Project. CSIRO, Australia, p. 217pp.

DHI, 2014. Mike Flood - Urban, Coastal and Riverine Flooding. DHI Water & Environment, Denmark.

Gell, P., Haynes, D., 2005. A Palaeoecological Assessment of Water Quality Change in the Coorong, South Australia., A Report for the Department of Water, Land and Biodiversity Conservation. The University of Adelaide, Adelaide, South Australia.

Goverenment of South Australia, 2010. Water for Good, A Plan to Ensure Our Water Future for 2050. Report from the office of water security, Adelaide. ISBN 978-1-921528-34-7.

Heneker, T.M., 2010. Development of Flow Regimes to Manage Water Quality in the Lower Lakes, South Australia, DFW Technical Report 2010/05. Government of South Australia, through Department for Water, Adelaide.

MDBA, 2011. All about the Barrages, Licensed from the Murray–Darling Basin Authority under a Creative Commons Attribution 3.0 Australia Licence.

MDBA, 2013. Murray–Darling Basin Authority Annual Report 2012–13, Licensed from the Murray–Darling Basin Authority, under a Creative Commons Attribution 3.0 Australia Licence.

CHAPTER 12

A paradigm shift of water resource management from upstream reservoirs to downstream/coastal reservoirs in Malaysia

Sin Poh Lim[1], Fang Yenn Teo[2], Yew Chong Tan[3], Md Nasir Mohd Noh[4]
[1]Global Water Consultants Sdn Bhd, Kuala Lumpur, Malaysia; [2]University of Nottingham Malaysia, Semenyih, Malaysia; [3]Formerly from Ministry of Energy, Green Technology and Water, Putrajaya, Malaysia; [4]National Hydraulic Research Institute of Malaysia (NAHRIM), Malaysia

12.1 Introduction

Water is the core of sustainable development, where water scarcity affects more than 40% of the global population and this percentage is projected to rise. In Malaysia, despite having abundant annual rainfall, they still experience water stress in some major cities. To achieve Sustainable Development Goal (SDG) 6, the Malaysian government recognizes the need to harvest water using alternative methods. As a result, Malaysia has recently seen a paradigm shift in water resources development works from traditional upstream reservoirs to downstream reservoir.

In the recent 2018 budget, the Malaysian government listed several water resources development projects emphasizing off-river storage (ORS), a downstream reservoir concept. This concept is gaining popularity after several successful ORS-implemented projects, solving both quantity and quality problems that persist in the traditional approaches of a runoff river scheme or regulated dam system. It has also proven to be more economically, socially and environmentally friendly (Unit Perancang Ekonomi, 2016-2020).

Malaysian water demands are anticipated to escalate over the next 20 years as the country continues to develop. The population is estimated to increase to 41.5 million by year 2040 as compared to the present population of about 30 million (2010 Population Census). This translates to an increase of 11.5 million population increase over the next 20 years. Fig. 12.1 shows the planning demand and water supply system development from 2010 until

Sustainable Water Resource Development Using Coastal Reservoirs
ISBN 978-0-12-818002-0
https://doi.org/10.1016/B978-0-12-818002-0.00012-5

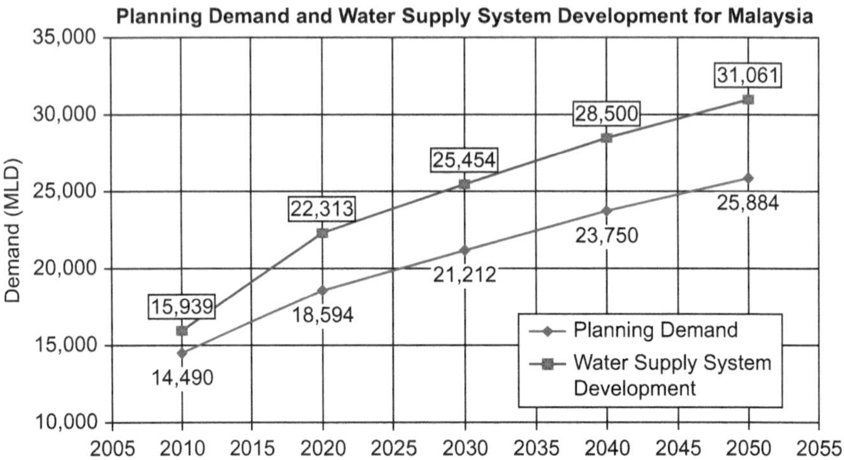

Figure 12.1 Long-term water demand and water supply development requirements for Malaysia.

Figure 12.2 Malaysia map indicating location of the major cities near the coast.

2050 (Department of Irrigation and Drainage, 2010). This indicates a need for water supply system development of about 10,000 MLd from 2010 to 2030. Considering the present stress on water resources in terms of quantity and quality in the major cities, this has become one of the key issues for the development in Malaysia. Fig. 12.2 shows the Malaysia map of the major cities located near coast. A sustainable solution to water resources development in this major city is critical to support the increasing water demand.

12.2 Evolution of water resources development in Malaysia

Currently, there are three main approaches to raw water abstraction in Malaysia as shown in Fig. 12.3. Surface water is the most extensively developed resource because of the abundance of rainfall in Malaysia, which

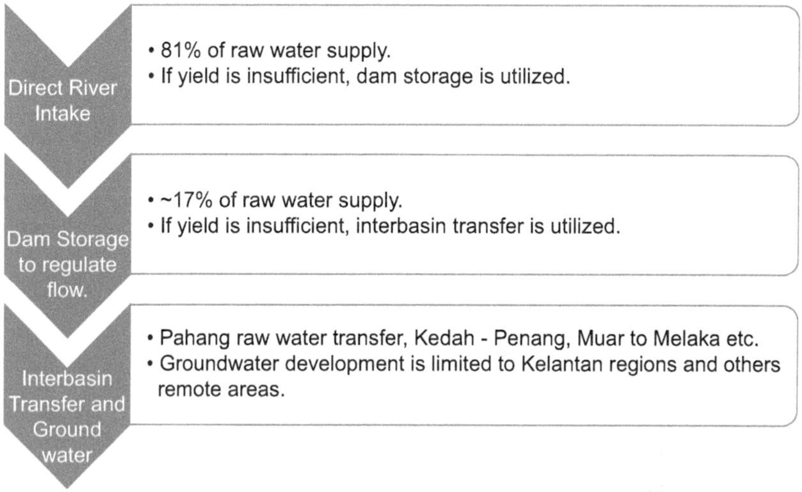

Figure 12.3 Previous water resources management approaches in Malaysia.

can be as high as 3310 mm annually at Terengganu (i.e., East Coast) and 3640 mm annually at Sarawak (i.e., East Malaysia). Fig. 12.4 shows the annual rainfall at each state in Malaysia. About 81% of raw water resources are directly abstracted from rivers for water treatment plants (WTPs). Flow regulation or direct supply by dam comprises 17%, while the remainder is supported by groundwater. Table 12.1 summarizes the issues and challenges of current management approaches.

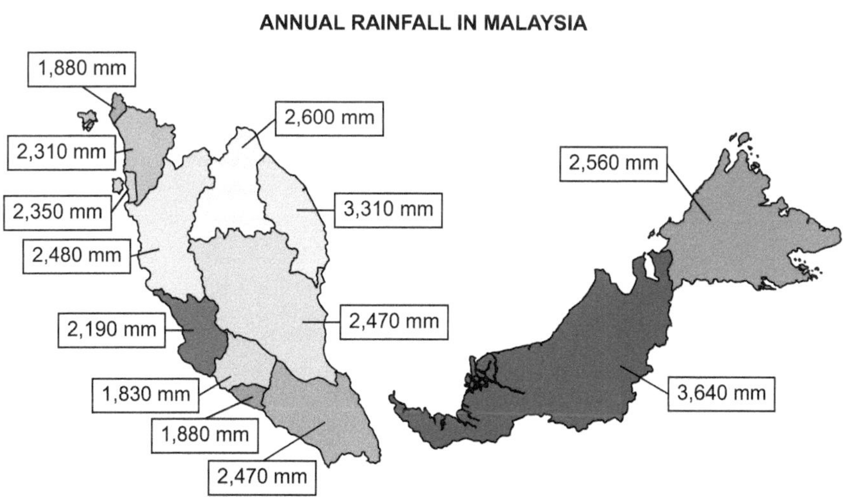

Figure 12.4 Malaysian annual rainfall distribution in each state.

Table 12.1 The issues and challenges of current approaches.

Approach	Challenges
1. Direct river intake 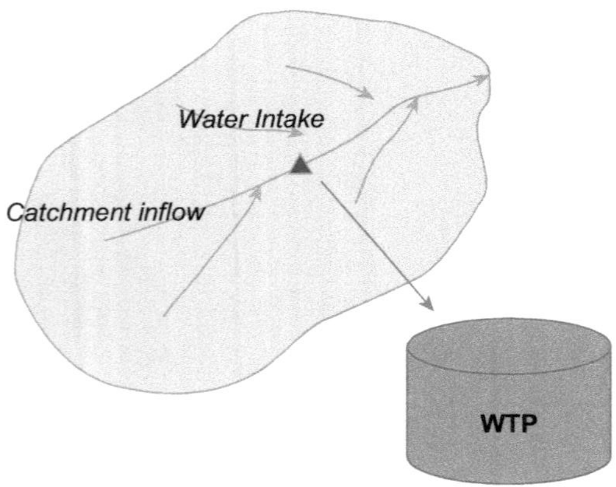	**(i)** Low river water levels during drought cause water to be unable to enter intakes and result in limited operation or shutdown of water treatment plants (WTPs). **(ii)** Sedimentation at river intakes requires dredging maintenance. **(iii)** Saline intrusion moves further upstream due to increase in upstream abstraction and sea level rise impacts. **(iv)** River pollution results in poor water quality, especially during dry seasons. Increasingly frequent exceedance of acceptable limits for WTPs, such as high ammonia content, has caused shutdowns. **(v)** High total suspended solids content results in high treatment cost. **(vi)** Lower river yields due to increase in upstream abstraction and climate change.

2. Regulating dam

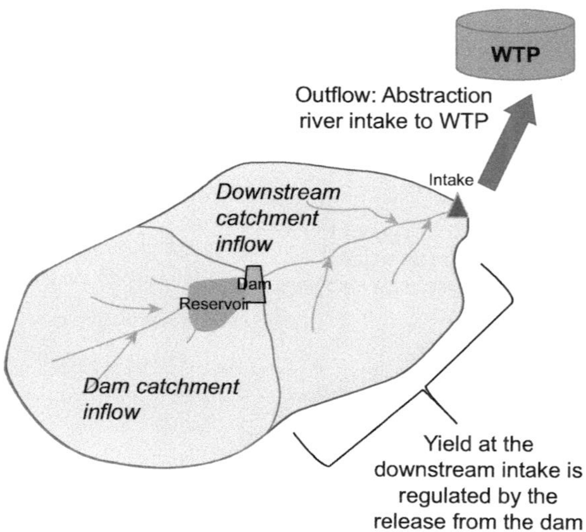

(i) Depleting reservoir water level during drought due to inland location of dams and reduced rainfall.
(ii) Active land development and agricultural activities upstream of the dam catchment cause reservoir water quality deterioration.
(iii) Illegal encroachment into dam reservoir areas or fringe developments.
(iv) Intake lies on a river; it will still face similar issues as 1 (i) to (v).

Continued

Table 12.1 The issues and challenges of current approaches.—cont'd

Approach	Challenges
3. Direct Supply dam 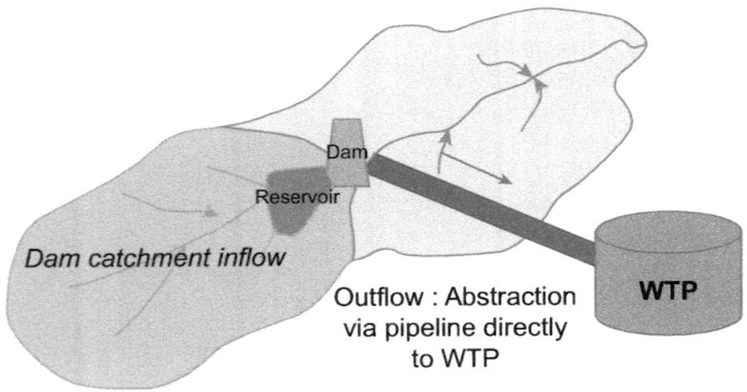	**(i)** Same challenges as 2 (i), 2(ii) and 2(iii). **(ii)** Extremely difficult to mitigate when reservoir water starts to deteriorate.
4. Groundwater 	**(i)** Development is unfavoured due to abundance of available surface runoff. **(ii)** Geologically dependent and more suitable for remote areas where water supply networks are not extensive. **(iii)** Involves high pumping and drilling costs. **(iv)** Inventory not available.
5. Interstate/intercatchment transfer	**(i)** Involved high capital expenditure (CAPEX) and operating expenses (OPEX). **(ii)** High pumping costs. **(iii)** Involves interstate contractual issues.

Runoff river schemes are facing both water quantity and quality problems. As more and more intakes are built along rivers to supply development in the same catchment, drought effects on flow and water levels become more pronounced. Intake yields are thus affected based on land use and development scenarios, and WTPs cease operations when river water quality does not meet required standards. High ammonia content is a frequently reported cause of disruption. WTPs at Cheras Batu 11, Bukit Tampoi, Skudai, Linggi and Sembrong in particular have been struggling with the river water pollution problems during dry season.

Water resources development by means of inland dam construction has become less favorable since its portrayal as a non—environmentally friendly solution, facing strong objections from the public and NGOs. Dam construction is counterproductive to government efforts and commitments in reducing the carbon footprint. State governments also express concerns of losing huge land areas, which are submerged underwater by the dam reservoir, as well as the constraints in developing the dam catchment postconstruction.

The uneven distribution of rainfall, particularly with relevance to the Western Coast urban centers where water demand is concentrated, has led to extensive interstate pipeline water transfers, which are known to suffer significant (i.e., as much as 50%) nonrevenue water losses over time. The Malaysian government embarked on the biggest interstate water transfer scheme in 2010 (i.e., the Pahang—Selangor raw water transfer scheme), which involved transferring of raw water from Sungai Semantan at Pahang through three diversion tunnels measuring 44.6 km in length to the Langat 2 WTPs at Selangor, completed recently. Other interstate transfer schemes currently under the planning stage include the Johor—Melaka, Melaka—Negeri Sembilan and Perak—Selangor water transfer scheme. These transfer schemes need to be reviewed as the energy cost involved during operation is high.

Fig. 12.5 summarizes policy shift to downstream storage water resources management approaches in Malaysia. In the interest of developing newer and more innovative water resources technologies at downstream, the state of Selangor government initiated the Hybrid Off-River Augmentation System (HORAS) in the year 2014. The scheme utilized the existing tin mining ponds located within the water catchment as storage to contain sufficient water for river flow regulation during the dry season. While HORAS managed to increase the water resources yield, there were concerns on the water quality because of the residuals of heavy metals within the beds of the ex-tin mining ponds.

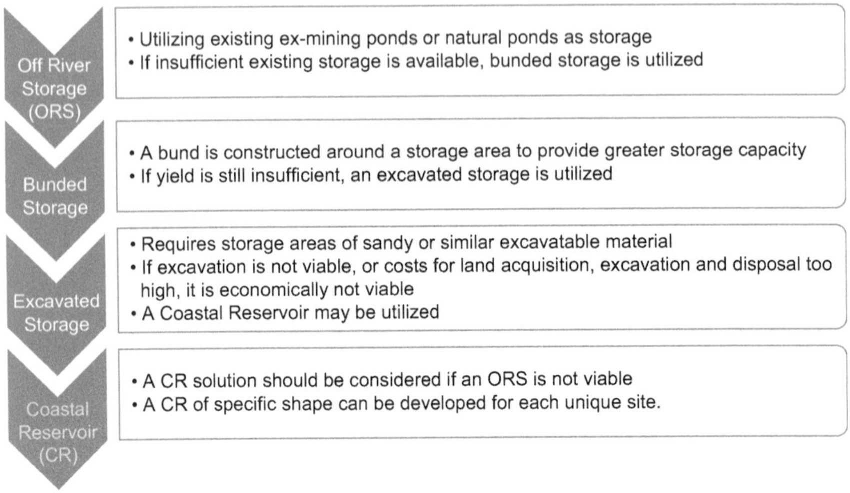

Figure 12.5 Policy shift to downstream storage water resources management approaches.

An alternative solution is needed to increase storage capacity without compromising water quality, and the downstream/coastal reservoir concept fits this purpose very well. The introduction of coastal reservoirs, promoted by the International Association of Coastal Reservoir Research (IACRR), provides another innovative solution in addition to ORS. While ORS utilizes existing pond storage facilities, the CR concept can be a good alternative if suitable pond sites for ORS have been exhausted. Compared to ORS which utilizes land space, CR creates additional space which can be potentially utilized not only for water resources but also for other purposes such as waterfront development, power generation and tourist spots. CR can be one of the sustainable solutions to solve the water scarcity problems in many coastal cities globally and in Malaysia, thus meeting the SDG 6 by the United Nations. Table 12.2 summarizes the approaches and challenges of new water resources management.

12.3 Research and development of coastal reservoir as an innovative and sustainable solution

12.3.1 Water availability in Malaysia

Based on the rainfall depths and recipient land surface area in Malaysia, there is an annual rainfall volume of roughly 971 billion m^3. After the assumption that 50% of rainfall translates to surface runoff, factoring

Table 12.2 Approaches and challenges of policy shifts to downstream water resources management.

Approach	Challenges/Remark
6. Off-river Storage 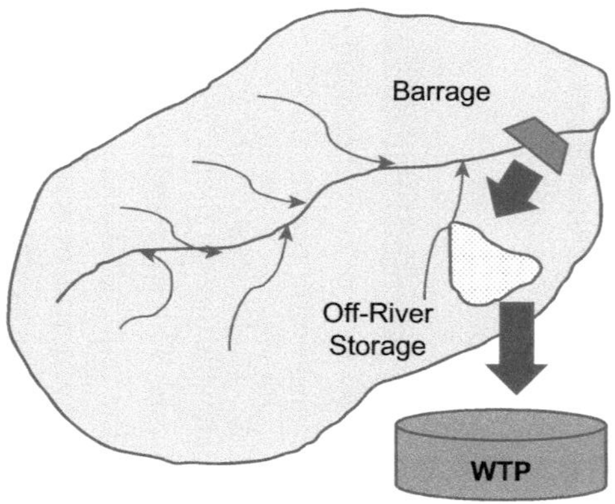 	**(i)** Utilizing an existing ex-mining pond or natural pond to store water to regulate flow. This is viable as it utilizes the water storage and does not involve changes in the water surface area. **(ii)** When ex-mining and natural ponds are exhausted, dredging to form new ponds involved changes from land surface to water surface areas. Disadvantages **(i)** May potentially create new environmental problems because of changes of land cover to water surface. **(ii)** Loss of land areas for land development. **(iii)** High land acquisition cost. **(iv)** High dredging and disposal costs (reduced if the soil is of sand that can be mined).
7. Downstream/coastal Reservoir 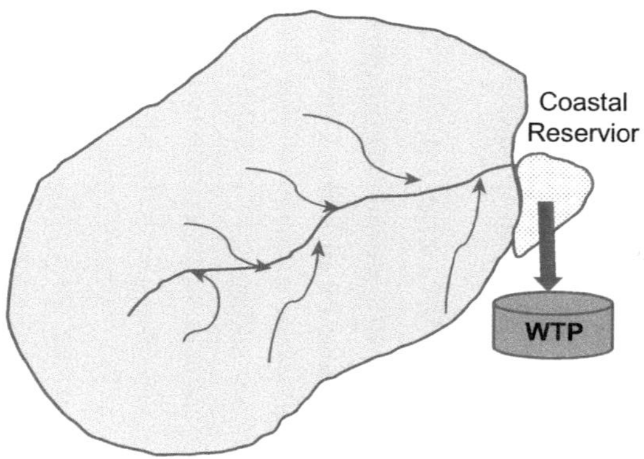 	**(i)** New approach for Malaysia. New approaches often need more time for acceptance. **(ii)** CR configuration is site-specific. Requires detailed study and planning before implementation. Specialist input is required. **(iii)** Main issues in planning, study and design are saline intrusion, water quality and costing.

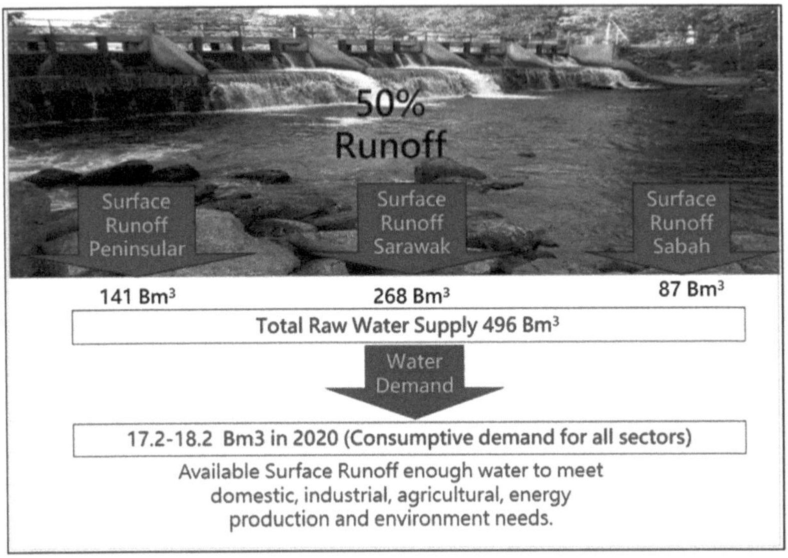

Figure 12.6 Water availability in Malaysia.

evaporation losses and groundwater recharge, the surface runoff volume is approximately 496 billion m^3. The total water demand as of 2020 is about 18.2 Bm^3 for all sectors, which is only 4% of this volume; see Fig. 12.6.

As such, Malaysia is not running out of water but water is running out of Malaysia. Downstream/coastal reservoir technology can be a potential long-term and sustainable solution for water resources in Malaysia. The present river runoff and dam system utilizes about only 3.7% of the available water, with most of the remaining water discharged into the sea. Water shortage is primarily an issue of storage rather than availability.

12.3.2 Why coastal reservoir?

Instead of existing mining ponds, storage can be created downstream or nearshore by utilizing a river reservoir, oxbow lake or by constructing the reservoir at nearshore (refer Table 12.2). Considering the demand increase of an additional 10,000 MLd in the next 20 years, with provision of 3 months storage, this will require a total volume of about 900 MCM or not even 1% of the total raw water available. This suggests that utilization of a downstream/coastal reservoir to store the required water is sufficient to meet the long-term water demand in Malaysia and to resolve the water shortage problem during drought by storing the excess water during wet seasons.

Traditionally, dams and reservoirs are sited in the upper catchment, river intakes are located at the middle catchment to avoid saline intrusion while demand centers are located typically at coastal regions. By locating the reservoir close to the demand center downstream, it harvests runoff from a bigger catchment, with much less environmental flow required and a much shorter water distribution pipeline. In contrast to conventional downstream river intakes, a downstream/coastal reservoir is also protected from salinity intrusion by virtue of its containing structure. A real-time water quality monitoring system allows only selected good quality water to enter the reservoir. Depending on the demand at a specific locality, water in the coastal reservoir can be utilized for various purposes such as domestic, irrigation or industrial usage.

12.4 Downstream/coastal reservoir as sustainable solution

SDGs are the 2030 Agenda for Sustainable Development adopted by the world leaders at the United Nations Sustainable Development Submit in September 2015, New York. SDGs cover a broad range of social and economic development issues, including poverty, hunger, health, education, climate change, gender equality, water, sanitation, energy, environment and social justice. It builds on the achievements and experiences of the Millennium Development Goals in year 2000, focussing on 17 goals and 168 targets. The SDGs are a global call to end poverty, protect the planet and ensure that all people enjoy peace and prosperity. SDG 6 is to ensure availability and sustainable management of water and sanitation for all. CR can be one of the sustainable solutions to ensure availability and sustainable water for Malaysia.

The World Commission on Environment and Development, 1987 defined 'sustainable' as 'development that meets the needs of the present without compromising the ability of future generations to meet their own needs'. Based on this interpretation, engineering solutions must be environmentally friendly, economically viable and socially acceptable. Environmentally, downstream/coastal reservoirs have limited local impact compared with inland dams and reservoirs. It can be designed to minimize environmental impacts, as illustrated by the Shanghai Qingcaosha reservoir taking advantage of the existing river alignment and geographical condition (refer Fig. 12.7). Located at the river mouth, nearshore or alongshore, CR is capable of providing extensive water storage capacity as it is not limited by

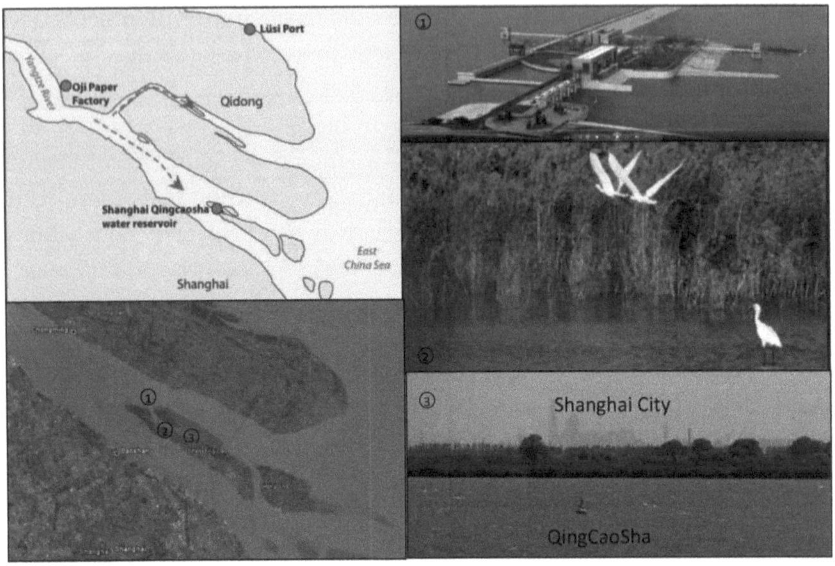

Figure 12.7 Qingcaosha reservoir maintains freshwater storage without obstructing waterways.

land area. The reservoir is filled by either passing freshwater through a gate system during river flow or flooding or by way of a conventional catchment system where the downstream/coastal reservoir is fed directly by catchment runoff. Downstream/coastal reservoirs are designed in such a way that they can be adapted to different locations without blocking off entire waterways as shown in Fig. 12.7 and thus do not disturb environmental flows or require rerouting of channels.

The retaining structures of a downstream/coastal reservoir can be constructed of concrete, earth or other materials depending on the soil condition. The primary barrier should be high enough to avoid tidal influx and significant wave height and be able to withstand the forces imparted on the wall by wave and tidal actions. The Qingcaosha reservoir was formed by a dike structure with a depth of about 25 m, which is a comparatively lower risk structure compared with conventional dam heights. The relative construction costs are economically reduced by utilizing local sand material as part of the dike body. A freshwater reservoir such as Qingcaosha also creates a man-made wetland making it environmentally and aesthetically friendly in comparison with traditional dam construction. Fig. 12.7 shows the constructed wetland at Qingcaosha reservoir.

12.4.1 Ensuring good water quality for coastal reservoir

Water quality is critical for downstream coastal reservoir. However, the Qingcaosha reservoir has successfully addressed this problem by the following methodologies:

- Installation of series of real-time water quality monitoring station at upstream of the inlet structures. By doing so, the inlet gate will only open when the water quality at inlet met the required water quality standard. Therefore, a system can be designed to allow only selective good water quality water to enter the reservoir.
- Construction of wetland at the upper part of reservoir manages to function as natural filtration system to improve the water quality. The same concept can be adopted for CR. In Malaysia, Putrajaya constructed wetland is one of the success stories of utilizing wetland to improve the water quality in the Putrajaya Lake. Fig. 12.8 shows the Putrajaya wetland system, the concept and site conditions. Water quality at Putrajaya Lake managed to achieved Class IIA water quality by having the upstream wetland system (PLWMOS).

Malaysian monsoon patterns of weather, with seasonal rainfall patterns and monthly distributions (refer Fig. 12.9), make it more conducive for CR implementation. The highest rainfall typically falls within the end of the year, while dry seasons are typically about 2–3 months in which water shortages emerge. Storage of excess flood water during wet seasons allows for distribution during the next dry spell. Sizing of the storage required at the downstream/coastal reservoir can be estimated by assuming zero inflow into the reservoir during drought for worst-case scenario. Actual storage requirements can be further refined during detailed design and detailed reservoir storage simulations.

12.5 Advantages of downstream/coastal reservoir

In comparison with traditional inland dams located upstream in Malaysia, downstream/coastal reservoirs provide the following advantages as shown in Table 12.3.

12.6 Way forward

Malaysia is a water-rich country. Improving water resources management by adopting the new and innovative approach of downstream/coastal reservoir can potentially increase the utilization of raw water resources. For Malaysia to meet water demands, increasing the utilization of surface runoff from the present 3.7%–5% will be sufficient. This potentially resolves the

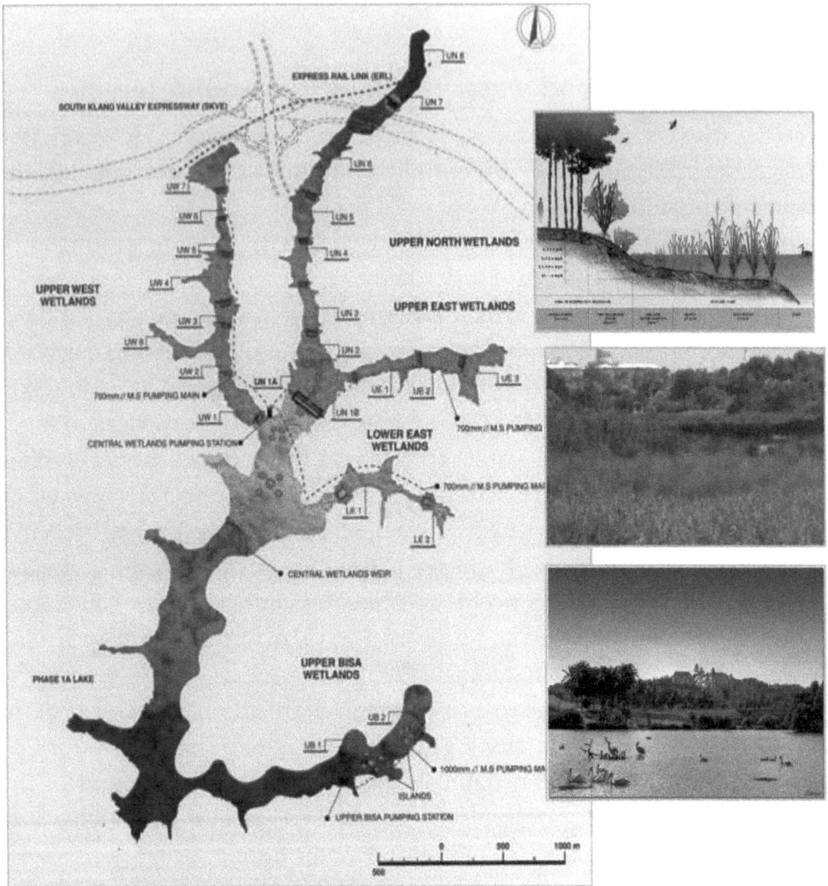

(Source: PLWMOS, Putrajaya Lake and Wetland Management & Operational System)

Figure 12.8 Putrajaya wetland system with Class IIA water quality at downstream Putrajaya Lake.

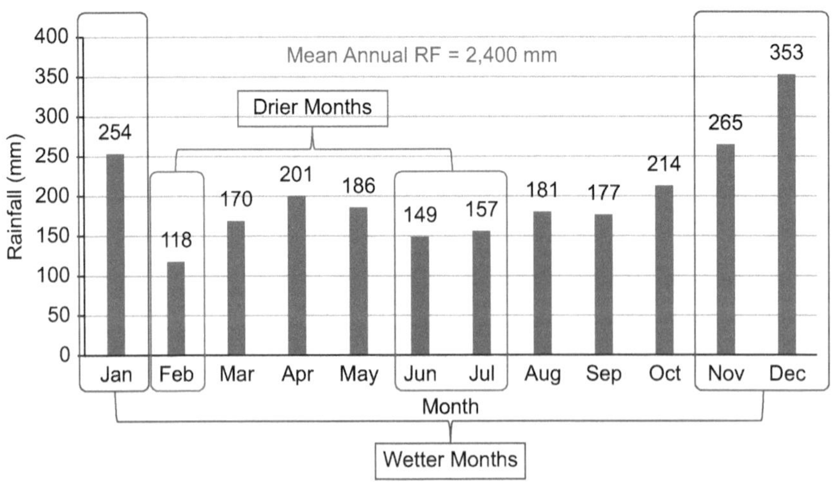

Figure 12.9 Typical monthly rainfall distribution of Malaysia.

Table 12.3 Comparison of downstream/coastal reservoirs against conventional dams.

Comparison	Dam	Downstream/Coastal reservoir
1. Land acquisition	A large land area is required to be inundated	Minimum land acquisition, utilizing river reserves, buffers, waterways, shore area, etc.
	Loss of productive land	Potentially creates new land area and enhances surrounding property value with a significant waterbody
2. Environmental Impact	Loss of fauna and flora	Create man-made wetland and new ecosystem
	Loss of green area and thus carbon absorption	Minimum impact
3. Social impact	Creates social issues and faces strong objection from locals and public	Less social issues as site selection can be very flexible
4. Heritage/ historical site	May inundate heritage or historical site	Can be avoided as site selection is flexible
5. Distance to demand point	- Very far, can be up to hundreds of km - High energy cost as it involves booster pumping from sources to demand points - More losses due to longer pipe length	Site near demand points and low energy cost Less losses due to much shorter pipe length
6. Catchment area	- Often sited at upstream, thus having smaller catchment area - During drought, reservoir gathers no rainfall	Located downstream, thus having much larger catchment area, gathers flow even during low flow
7. Expandability	Limited and difficult	Can be easily expanded
8. Risk	Create dam break risk to downstream population and properties	Low risk
9. Construction	Difficult and slower	Simpler and faster
10. Maintenance cost	Higher	Lower
11. Lifespan	Limited	Longer

water shortage problem during drought by storing the excess water during wet seasons. There are many significant advantages to adopting downstream/coastal reservoirs as compared with inland dam reservoirs or other alternate solutions such as desalination plant, which is not necessary for Malaysia with its abundant raw water resources. CR is overall a cost-effective, environmentally friendly, green and sustainable solution for raw water resources development in Malaysia.

To expedite the implementation of the CR system, the State Government of Malaysia is recommended to consider the construction of CR for major coastal cities in the country such as Johor Bahru, Melaka, Penang, Kuantan and Bintulu. Setting CR as one of the requirements for new onshore, nearshore or reclamation land development at these areas can be one of the solutions.

CR planning and future operations involve a lot of research in topics such as saline intrusion and water quality. The establishment of CR research centres at local universities is strongly recommended. Support from government agencies in terms of R&D is important to ensure the best implementation of the CR system in Malaysia for the near future. R&D on strategy and its applications in Malaysia is desirable to make sure of the successful implementation of CR. An integration on the science and engineering approaches needs to be highlighted for the best implementation of CR in Malaysia.

Acknowledgements

We acknowledge and thank the Ministry of Water, Land and Natural Resources, Malaysia and the Department of Irrigation and Drainage, Malaysia for their contributions in providing facts and figures for this paper.

References

Department of Irrigation and Drainage, 2010. Review of the National Water Resources Study (2000-2050) and Formulation of National Water Resources Policy.

PLWMOS, Putrajaya Lake and Wetland Management & Operational System.

Unit Perancang Ekonomi, RANCANGAN MALAYSIA KESEBELAS 2016-2020, PERTUMBUHAN BERPAKSIKAN RAKYAT, ISBN 978-9675842078, Percetakan Nasional Malaysia Berhad.

CHAPTER 13

Remote sensing image based nearshore bathymetry extraction of Mangaluru coast for planning coastal reservoir

Prerna Kumari, H. Ramesh
Department of Applied Mechanics and Hydraulics, National Institute of Technology Karnataka, Surathkal, Mangaluru, Karnataka, India

13.1 Introduction

Most of the world's mega cities are situated on coastal plains as people continuously moving towards coastal regions due to many benefits, and as a result many more such mega cities will form in the near future. This significantly increases the pressure on the most important resource called 'domestic water supply' in coastal region and aquifers (Herrera-León et al., 2018). Due to changes in climate and anthropogenic activities, rainfall is not uniformly distributed but in recent years many countries have experienced extremes either of cyclones or droughts. This could alter supply of water against demand. So to cope with future water needs, construction of coastal reservoirs are necessary, which have benefits including less environmental damage, replacement of villages, submergence of forest, etc. as compared to inland reservoirs. There are number of coastal reservoirs (Yuan et al., 2007) that have been constructed in many countries such as China, South Korea, Singapore, and the United Kingdom in the recent decades. According to the literature, any reservoir that is constructed in the coastal environment, including nearshore, at the shore, bays, and across rivers just upstream of estuary, are called coastal reservoirs (Yang et al., 2013; Jin et al., 2019). To construct such coastal reservoirs, particularly in the nearshore region of sea, bathymetry is most important for selecting suitable site or location, minimizing resources and increasing the storage, and minimizing damage to the shoreline and marine ecosystem.

Bathymetry is called the study of beds or floors of water bodies, including rivers, streams, sea or oceans, lake, etc. It originally refers to the depth of ocean relative to sea level, in other words, called submarine

Sustainable Water Resource Development Using Coastal Reservoirs
ISBN 978-0-12-818002-0
https://doi.org/10.1016/B978-0-12-818002-0.00013-7

topography, meaning the depth and shape of underwater land. Like topography that gives the three-dimensional representation of overland terrain, bathymetry also gives the representation of underwater terrain, and holds a very important role in the hydrographic study of any water body. Bathymetric surveys allow to measure the depth of a water body as well as to map the underwater features. Multiple methods can be used for bathymetric surveys including multi-beam and single beam echo sounder surveys, Acoustic Doppler Current Profiler (ADCPs), sub-bottom profilers, and the Eco-mapper Autonomous Underwater Vehicle, which was advanced by light detection and ranging (LIDAR) datasets. Bathymetric surveys are used for many research fields including flood inundation, the contour of streams and reservoirs, leakage, scour and stabilization, water-quality studies, dam removal, biological studies storage in reservoirs and ponds, planning the coastal reservoir, etc.

Remote-sensing techniques have provided an advanced, easy, and economical method for shallow water depth estimation, because of availability of multi-temporal, multi-resolution, and multi-constellation satellite data freely such as MODIS/TERRA (Clark et al., 1987; Ayana et al., 2015; Ovakoglou et al., 2016). Time-different methods have been proposed for shallow water depth estimation, including SONAR soundings and LIDAR data, photogrammetric-based bathymetry (Hodl et al., 2018), which is an expensive source of data and is area restricted. Also, it is difficult for acquiring ship soundings at very shallow depths and area with coral reefs, etc. A traditional ship sounding method needs a proper set of expensive equipment for carrying out the work, which increases the overall cost of the project and is also restricted to very shallow areas. To overcome this problem of ship soundings, LIDAR (Minzhang et al., 2014) is used, which is an optical remote-sensing technique that uses laser light to densely sample the surface of the earth, producing highly accurate *x, y, and z* measurements. Airplanes and helicopters are the most commonly used platforms for acquiring LIDAR data over broad areas. But LIDAR data collection over an area is also a high-expenditure job. To overcome the difficulties and disadvantages of these traditional methods, satellite-based remote-sensing techniques have been developed over the years for better estimation of shallow depths (Pacheco et al., 2015). Altimetry data has been used as an alternative to these techniques (Jena et al., 2012). Another method is using algorithms for the estimation of nearshore bathymetry, called satellite-derived bathymetry (SDB) (Smith and Sandwell, 2004).

SDB refers to the depths processed from optical satellite imagery. SDB is based on the expectation that deeper water appears darker than shallower water. The principal concepts driving SDB first came about in the 1970s when multispectral satellite sensors such as those aboard the Landsat satellite began orbiting the Earth (Kent et al., 1988). Multispectral sensors collect specific wavelengths of energy along the electromagnetic spectrum, referred to as bands. Typically, sensors collect reflected bands of visible light like red, green, and blue, and may include other non-visible reflected and emitted bands such as near and far infrared. In general, SDB data can offer:

1. Good coverage (within depth and image limitations); not as good as multi-beam echo sounder (MBES), some objects may be missed, but better than single-beam echo sounders (SBES) and lead line.
2. Better object detection than lead line, but not as good as SBES used with side scan sonar or an MBES.
3. Good positional accuracy; similar to MBES and SBES; better than historic lead line.
4. Lesser depth accuracy than MBES, SBES, and lead line.

At the core, SDB is the connection between the reflected energy and water depth. Essentially, for each pixel of a given satellite image where the sea floor is visible, there is a statistical relationship between the amount and type of energy that the sensor detects, and the depth of the water at that pixel location. This relationship can be exploited with computational algorithms. Complex mathematical analysis (Fotheringham et al., 1998) of the imagery attempts to distinguish these differences between deep-water pixels and anything appearing darker in the image and remove many other sources of noise in a satellite image and produce the best estimate for the average depth in each pixel. Satellite imagery is available at different spatial resolutions (+100 m to 31 cm). Only imagery of a resolution higher than 30 m is suitable and appropriate for charting (Bailly, 2011), as SDB results in an average depth per pixel. Even when using 30 m resolution imagery, the results should be used with caution as many shallow depths may not have been detected. SDB processing requires multispectral imagery, which in the commercial sector is currently limited to a maximum resolution of 1.24 m (WorldView-3). Higher resolution imagery is usually black and white and not able to provide SDB. Unlike active depth measurement techniques such as echo sounders or LIDAR, where controlled signals are transmitted and received, SDB is a passive technology and is simply measuring the reflected sunlight intensity. Because of this, SDB results are affected by many more uncontrollable environmental factors. SDB is limited to shallow clear waters where the seabed can be seen in the imagery. Its results can be adjusted and improved by providing ground truth data using more conventional survey techniques (e.g., echo sounder). SDB can be implemented through either analytical or empirical methods.

Empirical methods explore the statistical relationships between image pixel values and field-measured water depths. Analytical approaches rely on the general principle that the seawater transmittances at near-visible wavelengths are functions of a general optical equation dependent on the intrinsic optical properties of seawater. A number of external factors affect the accuracy of the depth calculation, including the spatial and spectral resolution of the imagery, the viewing angle of the satellite, the solar illumination angle, atmospheric effects, sunlight, tide level, and submerged vegetation. Careful selection of satellite imagery and subsequent image processing can mitigate some of these effects. This method has been a topic of interest due to its easy application and free availability of data. Different algorithms have been developed in the past 30 years to determine the nearshore bathymetry or shallow-water depth estimation (Holman et al., 2013; Vinayaraj et al., 2015). The ability of light to penetrate the water body provides a physical basis for modelling water depth from remote-sensing spectral data (Deng et al., 2008). Different SDB methods fall into the category of spectral rationing and radiative transfer models. In case of radiative transfer, single-spectral band and multispectral band models have been proposed. The single-band algorithms assume a constant attenuation coefficient and homogeneous bottom type. Reliable SDB is possible when the water is clear and when water quality and bottom types are homogeneous. When such conditions are satisfied, single-band water depth models can provide a reasonable estimate of depth. Nonetheless, coastal water environments rarely offer such ideal conditions. Depth estimates from satellite depend upon the amount of light received at the sensor, which is a function of atmospheric conditions, depth, in water volume scattering, etc. The depth obtained using reflectance at the sensor is highly affected by different parameters such as atmospheric, water quality, bottom albedo, sediments present in the water and their concentrations, etc. Some other factors like the angle of the Sun, azimuth, height of platform, as well as scatter, absorption, and refraction in the atmosphere might affect the characteristics of the returned electromagnetic waves.

Most established SDB algorithms (Jawak et al., 2015) require some registration with known depth points. These registration points help calibrate the algorithm to the imagery by giving the computation a frame of reference. This tells the computer that when the seafloor is at a certain depth, specific types and intensities of energy (light) are collected by the sensor. Algorithms also utilize indices of known sea-bottom types and their specific reflectance signatures to further refine accuracy. While it may stand to reason that deeper water is darker, some seafloor types like seagrass or coral can reflect light that throws sensors off, creating erroneous depth values Fig. 13.1. Bathymetric information on lakes and reservoirs holds an important value in hydrology. Besides water level, volume-lake area or stage curve relationships, the multi-temporal comparison between bathymetries is an indicator of environmental changes like the lake or reservoir sedimentation. From this information, reservoir ecosystem functioning, lifetimes of reservoirs or erosion-sedimentation rates of catchments can be derived.

Monitoring reservoir/lake bathymetry has become attractive using recent advances in global positioning systems (GPS), portable sonar sounders, remote sensing data, and GIS software analysis tools. This study is an application of remotely sensed image and numerical algorithms to determine the bathymetry of the nearshore of Mangaluru coast, India, for planning a coastal reservoir for sustained water supply for Mangaluru city and nearby villages. The excess flood water from the Nethravathi river will

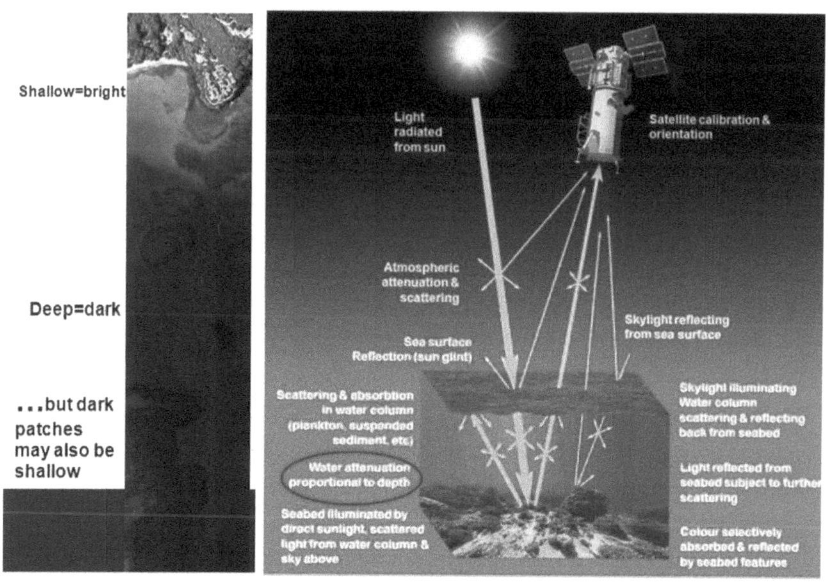

Figure 13.1 SDB methodology with signatures. *(Source: Dickerson, D., UK Hydrographic Centre.)*

be diverted without blocking the natural regime to this coastal reservoir if the quality of flood water is above standards, particularly suspended solids, turbidity, and sediment load.

13.2 Materials and methods

13.2.1 Description of study area

Mangaluru city is a port city of Karnataka state, India, with a city population alone estimated at about 622,557. Apart from this, the neighbouring villages contribute about 150,000. So totally around 0.8 million population is being supplied domestic water predominantly by treated surface water from the river Nethravathi and a very small portion by groundwater through open wells. The city experiences tropical humid climate. The average annual rainfall over this city is about 3500–4500 mm; more than 80% of this rainfall occurs during the southwest monsoon (June–September) and nearly 20% occurs in post-monsoon season. So having this much of annual rainfall, the city and nearby villages face acute scarcity of water for domestic use during summer months from March to May. The study area is the southwest coast of India (roughly 13°0′0″ N and 74°50′0″ E) Fig. 13.2, a low-energy wave environment with a tidal range of about 0.2–1.8 m.

13.2.2 Description of materials

The materials in the form of basic spatial and non-spatial data collected from remote-sensing platform and concerned organizations and used in this study are listed in Table 13.1.

LANDSAT 8 OLI/TIRS: LANDSAT 8 satellite sensor is part of the Landsat Data Continuity Mission that was successfully launched on 11 February 2013 from Space Launch Complex-3, Vandenberg Air Force Base in California, USA. LANDSAT 8 satellite has two main sensors: the operational land imager (OLI) and the thermal infrared sensor (TIRS). OLI collects images using nine spectral bands in different wavelengths of visible, near-infrared, and shortwave light to observe a 185 km (115 miles) wide swath of the Earth in 15–30 m resolution covering wide areas of the Earth's landscape while providing sufficient resolution to distinguish features like urban centres, farms, forests, and other land uses.

Hydrographic charts: Hydrographic charts for 2017–18, having the horizontal scale of 1:2500, horizontal datum of WGS 84, and projected to UTM Zone 43N was used in the study. These hydrographic charts are the

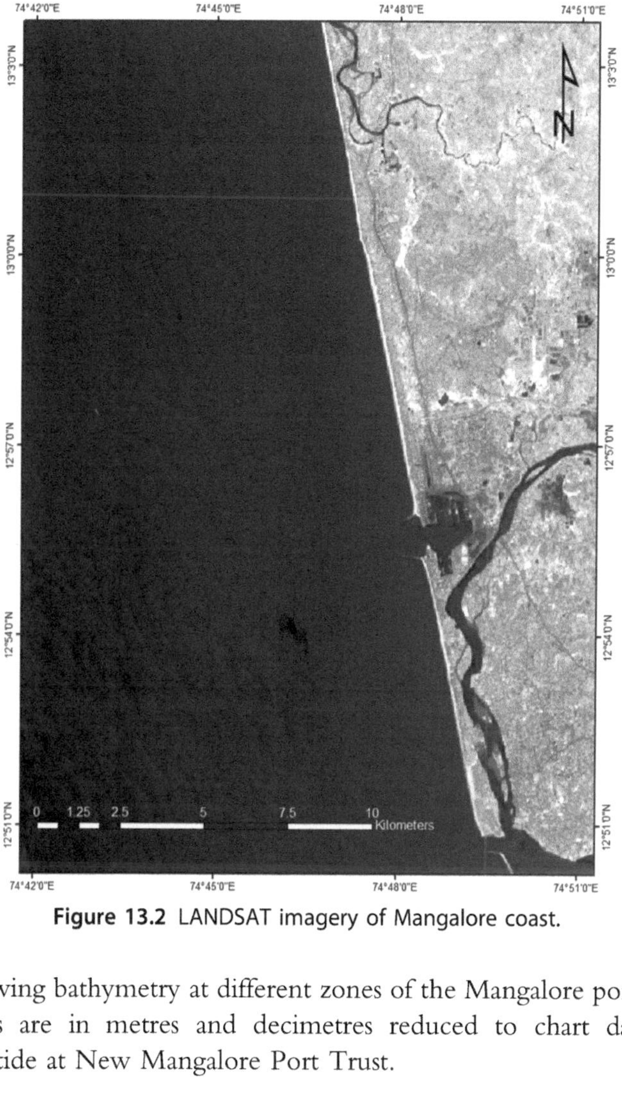

Figure 13.2 LANDSAT imagery of Mangalore coast.

maps showing bathymetry at different zones of the Mangalore port region. Soundings are in metres and decimetres reduced to chart data using observed tide at New Mangalore Port Trust.

13.2.3 Methods

In the present study, linear regression techniques (Brunsdon et al., 1996; Gollini et al., 2015; Vinayaraj et al., 2016) were used, which refers to a group of techniques for fitting and studying the straight-line relationship between two variables. Linear regression estimates the regression coefficients β_o and β_1 in Eq. (13.1):

$$Y_j = \beta_0 + \beta_1 {*} X_j + \varepsilon_j \tag{13.1}$$

Table 13.1 List of data used.

Data	Date of acquisition	Source	Description
LANDSAT 8 OLI/TIRS imageries	24 March 2017	USGS Earth Explorer earthexplorer.usgs.gov	Multispectral (11 bands) dataset, with a resolution of 15 m (Panchromatic) and 30 m(others)
LANDSAT 8 OLI/TIRS imageries	01 March 2017	USGS Earth Explorer earthexplorer.usgs.gov	Multispectral (11 bands) dataset, with a resolution of 15 m (Panchromatic) and 30 m(others)
Hydrographic chart of Mangalore port ODA and EXT Lagoon	21 Sep. to 10 October 2017	New Mangalore Port Trust	Map showing bathymetry in the approach region ODA and EXT Lagoon
Hydrographic chart of Mangalore Port Zone I	21st to 23rd October 2017	New Mangalore Port Trust	Map showing bathymetry in the approach Zone I
Hydrographic chart of Mangalore Port Zone II	18th to 21st October 2017	New Mangalore Port Trust	Map showing bathymetry in the approach Zone II
Hydrographic chart of Mangalore Port Zone III	4th to 9th October 2017	New Mangalore Port Trust	Map showing bathymetry in the approach Zone III

where, X_j is the independent variable, Y_j is the dependent variable, β_0 is the Y intercept, β_1 is the slope, and ε_j is the error. These parameters are usually called as regression coefficients. The unobservable error component ε_j accounts for the failure of data to lie on the straight line and represents the difference between the true and observed realization of Y_i. There can be several reasons for such difference, e.g., the effect of all deleted variables in the model, variables may be qualitative, inherit randomness in the observations, etc. It assumes that ε_j is observed as independent and identically distributed random variable with mean zero and constant variance ε_j. Later, it will additionally assume that ε_j is normally distributed.

The simple and multilinear regression: The simple linear regression model (Kabiri, 2017) considers the modelling between the dependent and one independent variable. When there is only one independent variable in the linear regression model, the model is generally termed as simple linear regression model. When there are more than one independent variables in the model, then the linear model is termed as the multiple linear regression model.

The uses of linear regression analysis are:

1. To find an equation that describes or summarizes the relationship between two variables. This purpose makes the fewest assumptions. This is a popular reason for doing regression analysis. The analyst may have a theoretical relationship in mind, and the regression analysis will confirm this theory. Most likely, there is specific interest in the magnitudes and signs of the coefficients. Frequently, this purpose for regression overlaps with others.
2. To predict the response variable, such as sales, delivery time, efficiency, occupancy rate in a hospital, reaction yield in some chemical process, or strength of some metal, etc. These predictions may be very crucial in planning, monitoring, or evaluating some process or system. There are many assumptions and qualifications that must be made in this case, for instance, not to extrapolate beyond the range of the data. Also, interval estimates require that normality assumptions to hold. Control regression models may be used for monitoring and controlling a system, for example, to calibrate a measurement system or keep a response variable within certain guidelines. When a regression model is used for control purposes, the independent variable must be related to the dependent variable in a causal way. Furthermore, this functional relationship must continue over time. If it does not, continual modification of the model must occur.

The following assumptions must be considered when using linear regression analysis.

Linearity: Linear regression models are the straight-line relationship between Y and X. Any curvilinear relationship is ignored. This assumption is most easily evaluated by using a scatter plot. This should be done early in the analysis. Nonlinear patterns can also show up in residual plot. A lack-of-fit test is also provided.

Constant variance: The variance of the residuals is assumed to be constant for all values of X. This assumption can be detected by plotting the residuals versus the independent variable. If these residual plots show a rectangular shape, it can assume constant variance. On the other hand, if a residual plot shows an increasing or decreasing wedge or bowtie shape, non-constant variance (heteroscedasticity) exists and must be corrected. The corrective action for non-constant variance is to use weighted linear regression or to transform either Y or X in such a way that variance is more nearly constant. The most popular variance stabilizing transformation is to take the logarithm of Y.

Special causes: It is assumed that all special causes, outliers due to onetime situations, have been removed from the data. If not, they may cause non-constant variance, non-normality, or other problems with the regression model. The existence of outliers is detected by considering scatter plots of Y and X as well as the residuals versus X. Outliers show up as points that do not follow the general pattern. Normally when hypothesis tests and confidence limits are to be used, the residuals are assumed to follow the normal distribution. So independent residuals are assumed to be uncorrelated with one another, which implies that the Y's are also uncorrelated. This assumption can be violated in two ways: model misspecification or time-sequenced data:

1. Model misspecification. If an important independent variable is omitted or if an incorrect functional form is used, the residuals may not be independent. The solution to this dilemma is to find the proper functional form or to include the proper independent variables and use multiple regression.
2. Time-sequenced data. Whenever regression analysis is performed on data taken over time, the residuals may be correlated. This correlation among residuals is called *serial correlation*. Positive serial correlation means that the residual in time period j tends to have the same sign as the residual in time period $(j - k)$, where k is the lag in time periods. On the other hand, negative serial correlation means that the residual in time period j tends to have the opposite sign as the residual in time period $(j - k)$. The presence of serial correlation among the residuals has several negative impacts: (1) The regression coefficients remain unbiased, but they are no longer efficient, i.e., minimum variance estimates. (2) With positive serial correlation, the mean square error may be seriously underestimated. The impact of this is that the standard errors are under estimated, the *t*-tests are inflated (show significance when there is none), and the confidence intervals are shorter than they should be.

13.3 Methodology

Different algorithms have been developed in order to obtain shallow-water depth using multispectral datasets. SDB models have been purported to retrieve coastal sea bottom reflectance from satellite imagery and effectively utilize this information to generate coastal bathymetry (Baban, 1993; Su et al., 2014; Poliyapram et al., 2017). Researchers have investigated SDB algorithms over the last 30 years and proposed estimation methods falling into categories such as spectral rationing and radiative transfer models. In case of radiative transfer, single-spectral band and multispectral band models have been proposed. The single-band algorithms assume a constant attenuation coefficient and homogeneous bottom type. Reliable SDB is possible when the water is clear and when water quality and bottom types are homogeneous. When such conditions are satisfied, single-band water-depth models can provide a reasonable estimate of depth. Nonetheless, coastal water environments rarely offer such ideal conditions. Therefore, radiative-transfer models using linear regression of multispectral bands have yielded good results. In present work, radiative-transfer method has been used, i.e., single linear regression (SLR) equation, using only single band of the multispectral image to determine the depth and multi linear regression (MLR) equation, using two or more bands to determine the depth (Jagalingam et al., 2015). To achieve the objective of this study, the overall methodology followed is shown in Fig. 13.3.

13.3.1 Pre-processing data

Satellite imageries used for the study are shown in Table 13.1. The pre-processing steps conducted are shown in the workflow (Fig. 13.2), and described below. The pre-processing steps are the same for SLR and MLR methods.

13.3.2 Radiance conversion

The satellite images obtained from the LANDSAT are already geometrically and radiometrically calibrated. Digital numbers (DNs) were converted into physical units of band-averaged spectral units (watts/m^2 × strad × μm). The equation used for the conversion of DN values to at-sensor radiance is given by Eq. (13.2).

$$L = M_L{}^{*} Q_{Cal} + A_L \tag{13.2}$$

where, $L =$ TOA spectral radiance (Watts/(m^2 × srad × *ijm*)); $M_L =$ Band-specific multiplicative rescaling factor from the metadata (RADIANCE MULT BAND x, where x is the band number); $A_L =$ Band-specific additive rescaling factor from the metadata (RADIANCE ADD BAND

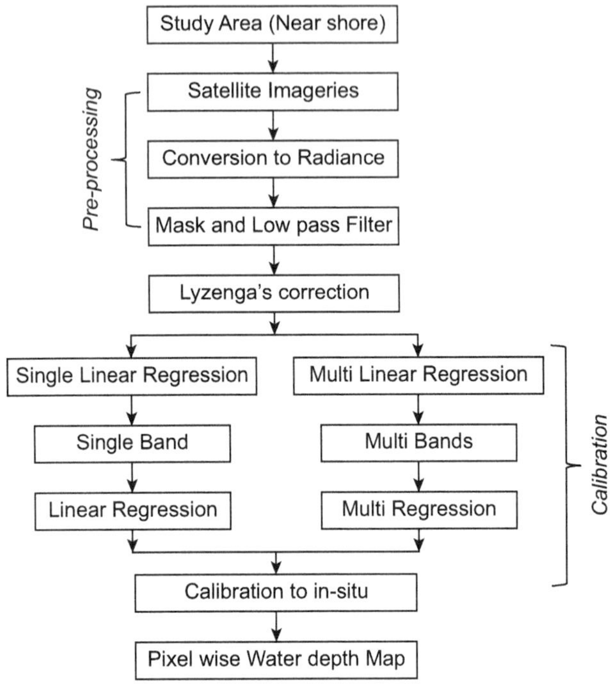

Figure 13.3 The overall methodology adopted.

x, where x is the band number) Q_{cal} = Quantized and calibrated standard product pixel values (DN).

13.3.3 Mask and low-pass filter

The very first step to carry out the study is to separate the water region from that of land. For doing so, the ratio of green to SWIR band is used as the ratio greater than one gives the water region and less than one gives the land region. To facilitate the study and for better understanding, the land region is masked from the image so that all the calculations are done only for the water regions. The image was corrected for smoothening to remove speckle noise by using 5×5 low-pass filter.

13.3.4 Lyzenga's correction

The radiance observed at the sensor is mainly composed of four components, viz., atmospheric scattering component, water surface reflectance component, in water volume scattering component, and the bottom reflectance component. The basic principle behind the determination of bathymetry from satellite data is the relation between the bottom reflectance collected at the sensor and the depth (Lyzenga et al. 2006; Kanno and Tanaka, 2012). So, it is necessary to remove all the three components not required from the radiance collected at the sensor prior to carrying on the further work. It is assumed that the deep-water pixels do not contain the bottom reflectance, and the radiance corresponding to the three other components is the same over the whole study area. Subtracting the average of deep-water pixel reflectance $L(\lambda\infty)$ from that of shallow-water pixels $L(\lambda)$ eliminates the other three components from the at-sensor radiance as given in Eq. (13.3).

$$C(\lambda)_i = L(\lambda)_i - L(\lambda\infty)_i \tag{13.3}$$

where, $C(\lambda)_i$ is the corrected radiance, i.e., radiance corresponding to bottom reflectance and 'i' corresponding to the respective band. Corrected radiance was calculated for blue, green, red, and NIR bands of LANDSAT data.

13.3.5 Depth estimation

Before application of the depth estimation algorithms, the corrected radiance was logarithmically transformed in order to obtain the relation between radiance and depth. It is given by the following Eq. (13.4):

$$X(\lambda)_i = Ln\left(C(\lambda)_i\right) \tag{13.4}$$

where i denotes the band used for analysis.

13.4 Results and discussion

13.4.1 The single linear regression method

The equation given by Stoffle and Halmo (1991), was used to determine the depth for the coastal region of Mangaluru (Eq. 13.5):

$$h = -\frac{1}{2}kX(\lambda)_i + \frac{1}{2}k(\log V_0) \tag{13.5}$$

where, h - estimated depth in metres, k - water attenuation coefficient, $1/2kX(\lambda)_i$ - the corrected transformed radiance, V_o is the sensitivity factor related to solar irradiance at surface, the bottom reflectance, atmospheric transmission, and sensor equipment (Gholamalifard et al., 2013). The bands used for the application of algorithm are blue and NIR, respectively. Fig. 13.4 shows the bathymetry generated using the respective bands. The range of depths from both the bands was obtained as 0–26 m in the selected region. The depth rasters were divided into five classes of depth,

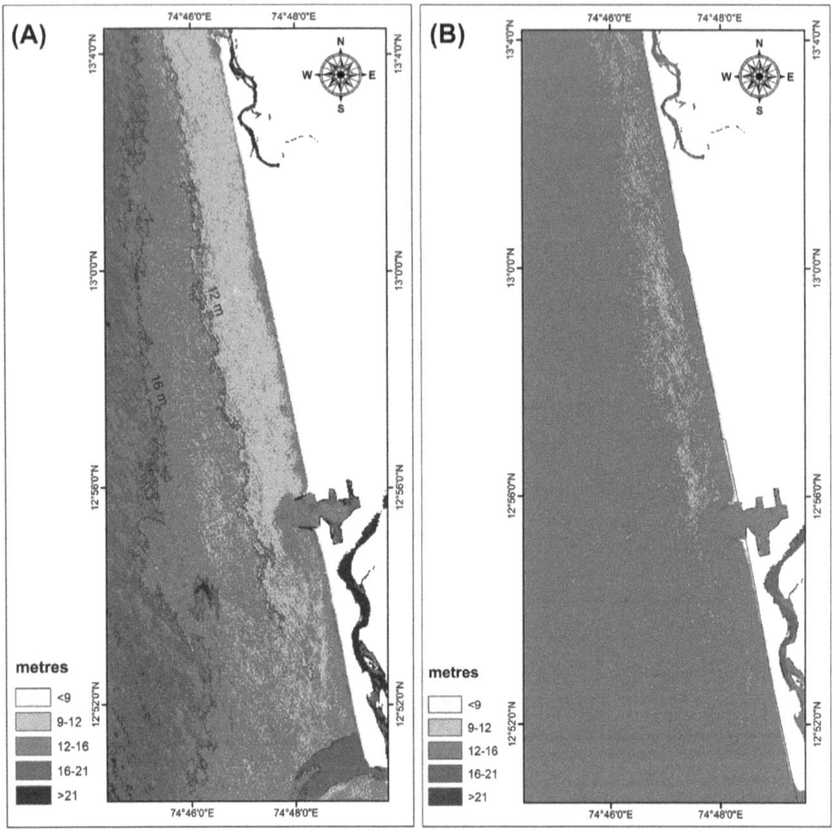

Figure 13.4 Bathymetry estimated from SLR method: (A) blue band and (B) NIR band.

ranging 0–9 m, 9–12 m, 12–16 m, 16–21 m, and greater than 21 m, as shown in Fig. 13.3. Though the ranges obtained from both bands were same, the blue band gave the more realistic estimation, while the NIR did not show the proper distribution of depth throughout the study area. Fig. 13.5 shows the scatter plot between ground truth depth and model estimated depth using (1) the NIR band, and (2) the blue band. The models showed correlation coefficient (r) of 0.87 and 0.84 and coefficient of determination (R^2) of 0.76 and 0.70, respectively.

13.4.2 Multiple linear regression method

It utilizes more than one band for regression analysis, i.e., blue, green, red, NIR etc. MLR is used so as to overcome the drawback of SLR of homogeneous bottom type. MLR does not consider the same bottom conditions and type for the whole study area as in the SLR method.

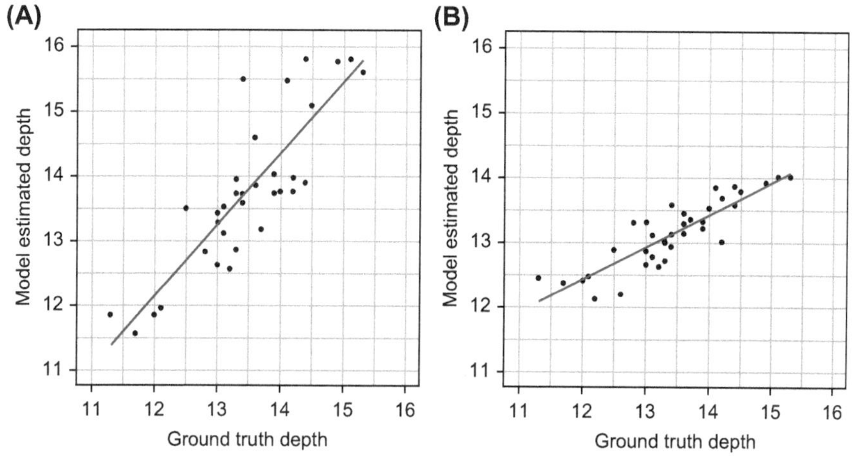

Figure 13.5 Scatter plot between ground truth and model estimated depth from SLR method: (A) blue band and (B) NIR band

Hence the method is more reliable and realistic. Eq. (13.6) given by Clark et al. (1987) is used for the analysis.

$$h=\beta_o+\beta_1X(\lambda)_1+\beta_2X(\lambda)_2+\ldots+\beta_nX(\lambda)_n \qquad (13.6)$$

The radiance value is calibrated against ground truth data to find the value of constants in the equation. Different band combinations were tried, but not all produced the accurate results. The first combination was using visible (i.e., Blue, Green, and Red) and NIR bands, from which the resultant depth map was not sufficiently accurate. So the next band combination used was Blue and NIR bands. The results of this combination were sufficiently accurate. The acquired depth map was also divided into the same classes, so as to identify the difference between the methods, as shown in Fig. 13.6. The accuracy of the algorithm was found to be higher than the SLR method as the correlation coefficient (r) between the model and the ground truth data is observed to be 0.94 and coefficient of determination (R^2) is observed as 0.88, which is considerably high. Also, the accuracy is affected by the resolution of the satellite imagery. Higher the resolution, higher the accuracy. The output bathymetry map is shown in Fig. 13.6A.

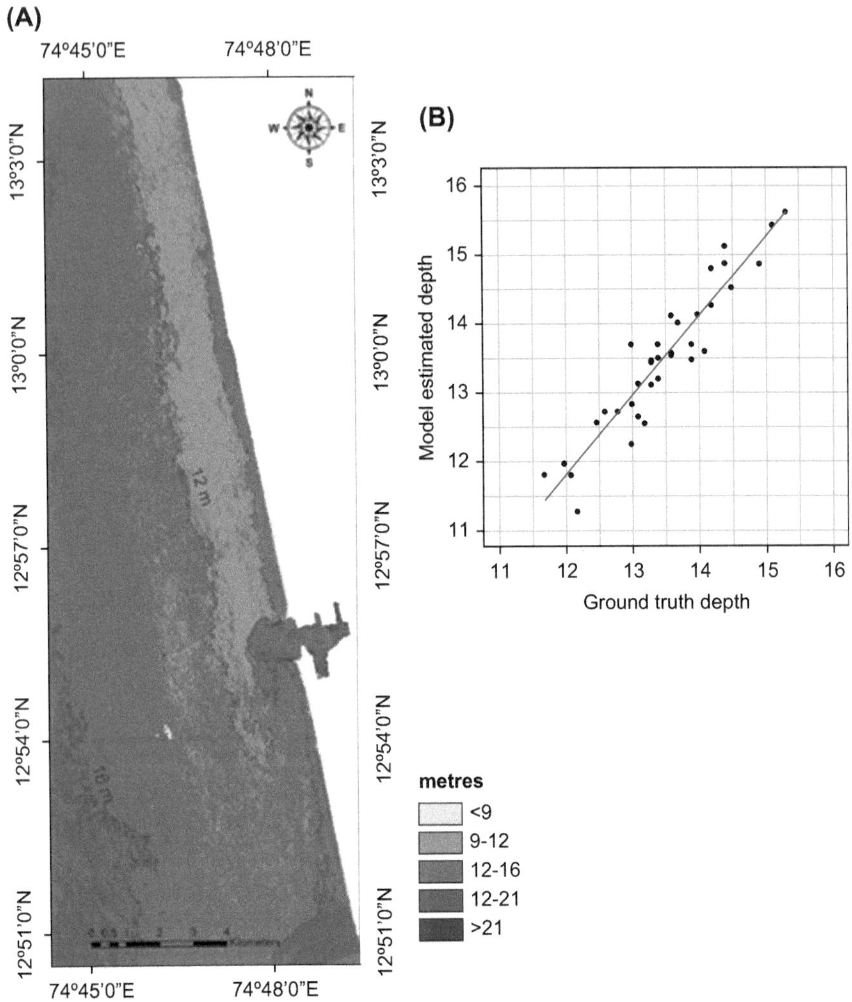

Figure 13.6 (A) Bathymetry of Mangalore coastal region using MLR method and (B) scatter plot of model estimated depth from blue and NIR bands from MLR method.

13.5 Summary and conclusion

Water is essential for survival of all living things and is been given top priority all over the world. Due to changes in the life styles of people and their activities, changes in climate alter both supply and demand for fresh water. So an alternate strategy for storing fresh flood water in the nearshore region of sea would be a future storage reservoir and called second-generation coastal reservoir. In order to plan such a coastal reservoir in the nearshore, water depth is very important because it provides the information for design of coastal reservoir dyke, the volume of water that is to be stored, materials required for construction, etc. So the present study has been conducted to extract bathymetry of Mangaluru coast using remote-sensing data. The study was carried out to apply the already-built algorithms for bathymetry determination for any water body. Single linear and multi linear regression equations have been used in this study. The study has been calibrated first for Mangalore port region against hydrographic charts. The correlation coefficient for the SLR method found to be 0.87 and 0.84, using single NIR and blue band, respectively. The coefficient of determination that resulted was 0.76 and 0.70, respectively. The accuracy from model was within a metre. But the drawback of the model is that the accuracy is not the same for the whole study area, because of changing bottom type. To overcome this problem, the MLR method was adopted, which gave the correlation coefficient of 0.94 and coefficient of determination of 0.88, which is considerably higher than the previous method. Hence MLR method was adopted to determine the bathymetry of the Mangaluru coast. From this bathymetry study, a suitable location for construction of coastal reservoir to store 2 thousand million cubic feet (TMC) of fresh flood water in the nearshore of Mangaluru coast could be possible within the depth of 0–16 m from the seashore with a suitable geometric shape of the dyke to minimize damage to seashore erosion and marine ecosystem and also help in assessing the materials and design of coastal reservoir dyke such as top width, bottom width, pressure computations, etc. Also, further impact studies need to be carried out on marine ecosystem and seashore dynamics through either numerical or physical models for the construction of coastal reservoir.

Acknowledgements

The authors are thankful to Science and Engineering Research Board (SERB), Department of Science and Technology (DST), Government of India for supporting this study as a part of the R & D project titled 'Impounding of River flood waters along Dakshina Kannada Coast: A sustainable strategy for water resource development' with FILE NO. IMP/2018/001298. The authors extend their thanks to New Mangalore Port Trust for providing hydrographic charts of the port area and all other satellite data providers such as NASA, USGS on open source platform.

References

Herrera-León, S., Lucay, F., Kraslawski, A., Cisternas, L.A., Gálvez, E.D., 2018. Optimization approach to designing water supply systems in non-coastal areas suffering from water scarcity. Water Resources Management 32 (7), 2457–2473. https://doi.org/10.1007/s11269-018-1939-z.

Ayana, E.K., Philpot, W.D., Melesse, A.M., Steenhuis, T.S., 2015. Assessing the potential of MODIS/Terra version 5 images to improve near shore lake bathymetric surveys. International Journal of Applied Earth Observation and Geoinformation 36, 13–21.

Baban, S.M.J., 1993. The evaluation of different algorithms for bathymetric charting of lakes using Landsat imagery. International Journal of Remote Sensing 14 (12), 2263–2273.

Bailly, P., 2011. Computers Geosciences Automatic calculation of bathymetry for coastal hydrodynamic models. Computers and Geosciences 37 (9), 1303–1310.

Brunsdon, C., Fotheringham, A.S., Charlton, M.E., 1996. Geographically weighted regression: a method for exploring spatial nonstationarity. Geographical Analysis 28 (4), 281–298.

Clark, R.K., Fay, T.H., Walker, C.L., 1987. Bathymetry calculations with Landsat 4 TM imagery under a generalized ratio assumption. Applied Optics 26, 4036–4038. The Optical Society (OSA).

Deng, Z., Ji, M., Zhang, Z., 2008. Mapping bathymetry from multi-source remote sensing images: a case study in the Beilun Estuary, Guangxi, China. The International Archives of the Photogrammetry, Remote Sensing and Spatial Information Sciences 37, 1321–1326.

Fotheringham, A.S., Charlton, M.E., Brunsdon, C., 1998. Geographically weighted regression: a natural evolution of the expansion method for spatial data analysis. Environment and Planning A 30 (11), 1905–1927.

Gholamalifard, M., Kutser, T., Esmaili-Sari, A., Abkar, A.A., Naimi, B., 2013. Remotely sensed empirical modeling of bathymetry in the south eastern Caspian Sea. Remote Sensing 5 (6), 2746–2762.

Gollini, I., Lu, B., Charlton, M., Brunsdon, C., Harris, P., 2015. GW model: an R package for exploring spatial heterogeneity using geographically weighted models. Journal of Statistical Software 63 (17), 50.

Hodl, M., Bird, S., Knudby, A., Chlnier, R., 2018. Satellite derived photogrammetric bathymetry. ISPRS Journal of Photogrammetry and Remote Sensing 142, 268–277.

Holman, R., Plant, N., Holland, T., 2013. cBathy: a robust algorithm for estimating nearshore bathymetry. Journal of Geophysical Research: Oceans 118, 2595–2609. https://doi.org/10.1002/jgrc.20199.

Jagalingam, P., Akshaya, B.J., Hegde, A.V., 2015. Bathymetry mapping using Landsat 8 satellite imagery. Procedia Engineering 116, 560–566.

Jawak, S.D., Luis, A.J., 2015. Spectral information analysis for the semiautomatic derivation of shallow lake bathymetry using high-resolution multispectral imagery: a case study of Antarctic coastal oasis. Aquatic Procedia 4, 1331–1338.

Jena, B., Kurian, P.J., Swain, D., Tyagi, A., Ravindra, R., 2012. Prediction of bathymetry from satellite altimeter based gravity in the Arabian Sea: mapping of two unnamed deep seamounts. International Journal of Applied Earth Observation and Geoinformation 16 (44), 1–4.

Jin, G., Mo, Y., Li, M., Tang, H., Qi, Y., Li, L., Barry, D.A., 2019. Desalinization and salinization: a review of major challenges for coastal reservoirs. Journal of Coastal Research 35 (3), 664–672. Coconut Creek (Florida), ISSN 0749-0208.

Kabiri, 2017. Discovering optimum method to extract depth information for nearshore coastal waters from sentinel-2a imagery- case study: Nayband bay, Iran. The International Archives of the Photogrammetry, Remote Sensing and Spatial Information Sciences XLII-4/W4, 105–110.

Kanno, A., Tanaka, Y., 2012. Modified Lyzenga method for estimating generalized coefficients of satellite-based predictor of shallow water depth. IEEE Geoscience and Remote Sensing Letters 9 (4), 715–719.

Kent, R., Fay, H., Temple, H., Mississippi, S., Charles, L., West, K., West, K., 1988. Imagery bathymetry using thematic mapper imagery. SPIE, the International Society for Optics and Photonics 925, 229–231.

Lyzenga, D.R., Malinas, N.P., Tanis, F.J., 2006. Multispectral bathymetry using a simple physically based algorithm. IEEE Transactions on Geoscience and Remote Sensing 44 (8), 2251–2259.

Minzhang, H., Jiancheng, L., Hui, L., Lelin, X., 2014. Bathymetry predicted from vertical gravity gradient anomalies and ship soundings 1 Introduction. Geodesy and Geodynamics 5 (1), 41–46.

Ovakoglou, G., Alexandridis, T.K., Crisman, T.L., Skoulikaris, C., Vergos, G.S., 2016. Use of MODIS satellite images for detailed lake morphometry: application to basins with large water level fluctuations. International Journal of Applied Earth Observation and Geoinformation 51, 37–46.

Pacheco, A., Horta, J., Loureiro, C., Ferreira, Ã.S., 2015. Retrieval of nearshore bathymetry from Landsat 8 images: a tool for coastal monitoring in shallow waters. Remote Sensing of Environment 159, 102–116.

Poliyapram, V., Raghavan, V., Metz, M., Delucchi, L., Masumoto, S., 2017. Implementation of algorithm for satellite-derived bathymetry using open source GIS and evaluation for tsunami simulation. ISPRS International Journal of Geo-Information 6 (3), 89.

Smith, W., Sandwell, D., 2004. Conventional bathymetry, bathymetry from space, and geodetic altimetry. Oceanography 17 (1), 8–23.

Stoffle, R.W., Halmo, D.B., 1991. Satellite Monitoring of Coastal Marine Ecosystems: A Case from the Dominican Republic. Consortium for Integrated Earth Science Information Network (CIESIN), Saginaw, Michigan.

Su, H., Liu, H., Wang, L., Filippi, A.M., Heyman, W.D., Beck, R.A., 2014. Geographically adaptive inversion model for improving bathymetric retrieval from satellite multispectral imagery. IEEE Transactions on Geoscience and Remote Sensing 52 (1), 465–476.

Vinayaraj, P., Raghavan, V., Masumoto, S., Glejin, J., 2015. Comparative evaluation and refinement of algorithm for water depth estimation using medium resolution remote sensing data. International Journal of Geoinformatics 11 (3), 17–29.

Vinayaraj, P., Raghavan, V., Masumoto, S., 2016. Satellite-derived bathymetry using adaptive geographically weighted regression model. Marine Geodesy 39 (6), 458–478.

Yang, S., Liu, J., Lin, P., Jiang, C., 2013. Coastal reservoir strategy and its applications. In: Wurbs, R. (Ed.), Water Resources Planning, Development and Management, pp. 95–115. USA.

Yuan, W.X., Yang, S.T., Zhuang, M., 2007. Arguments of the coastal reservoir in RuDong JiangSu Province. Yangtze River 38 (6), 35–37. https://doi.org/10.16232/j.cnki.1001-4179.2007.06.015.

CHAPTER 14

Future directions

Shu-Qing Yang, Muttucumaru Sivakumar
School of Civil, Mining and Environmental Engineering, University of Wollongong, Wollongong, NSW, Australia

14.1 Run-off to the sea is also one of the water resources to develop

Generally, people like to live by sea; this is why more than 54% of the world's population live within 200 km of the coast in 2001, and three-fourths of mega cities are situated by sea. Big pressure on land, water supply and other infrastructures appears in these coastal cities. In 2015, UN members endorsed the 2030 Agenda in which 17 grand sustainable development goals (SDGs) are listed. Among it, the SDG6 is about water supply and sanitation. All these 17 SDGs are interlinked, and water is the core. Without sustainable water supply, no country can achieve its SDGs. A big question is how to manage our water resources to meet our water demands for domestic, agricultural and industrial purposes. Among them, the water–food–energy nexus is especially critical.

It is interesting to note that the UN-Water only recommends the nature-based solutions for water; artificial water like desalinated water or wastewater recycling is excluded. Likewise, many countries such as China and Malaysia start to the integral management of land–sea resources. The International Association for Hydraulic and Environmental Research (IAHR) uses 'From snow to sea' as its main theme of next world congress. This implies that the world water leaders have realized that the current water solutions are invalid for SDG6 as these solutions are only limited in the land, and new solution must be extended to the sea. To make these intuitions more clear and sharpened, one has to resort to science first.

In the scientific communities, the definition of water resources is not very clear. A widely accepted definition is shown in Fig. 14.1, even the percentage may be different from this country to the other, but generally the floodwater and environmental flow are prohibited to develop. Consequently, in practice, only a small fraction of water is calculated as our utilizable water resources.

Sustainable Water Resource Development Using Coastal Reservoirs
ISBN 978-0-12-818002-0
https://doi.org/10.1016/B978-0-12-818002-0.00014-9

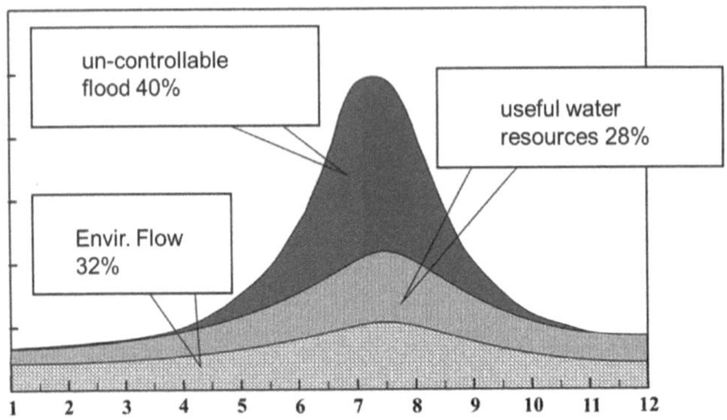

Figure 14.1 Conventional definition of water resources in 12 months (green portion: grey portion in printed version), in which the base flow or environmental flow and floodwater are not allowed to develop.

The broad definition of water resources is the surface water and underground water in the land. No matter which definition is used, it is common that the run-off to the sea is not the water resources. All assume that the run-off to the sea at a river mouth is too salty to drink; thus, desalination is needed for the coastal cities.

Different from this assumption, the International Association for Coastal Reservoir Research (IACRR) understands that the salinity at the river mouth (see Fig. 14.2) varies with the tide and river flow, i.e. not always seawater, at least during the flood seasons and low tide. Based on this, the IACRR claims that beyond the surface water and underground water, the run-off lost to the sea is also the water resources, or the third type of water resources. For SDG6, Fig. 14.2 shows that the total domestic water demand is 380 km^3/year, relative to the total runoff 45,500 km^3/year, only 0.8% of run-off is used for drinking purpose.

As the new definition of water resources is more scientific and realistic, one can see that the 50% of total population in the coastline can be fully met, and the thirst can be quenched. We take India as the example; now it is one of the most water-scarce countries and its 21 cities will run out of water in 2020. Its population in 2019 is 1.37 billion, and daily water consumption is 137 L/capita/day. Hence, the total Indian daily water consumption is 18,769,000 m^3, Indian river discharge is 63,789 m^3/s and its water volume to the sea in 50 min is higher than the total water demand. The same can be seen in other countries such as China, Malaysia and Australia. The driest continent in the world is Australia whose total run-off to the sea is about 440 km^3/year or 13952 m^3/s. Its population is 24 million in 2020, and the total daily water consumption is 32,88,000 m^3, equivalent

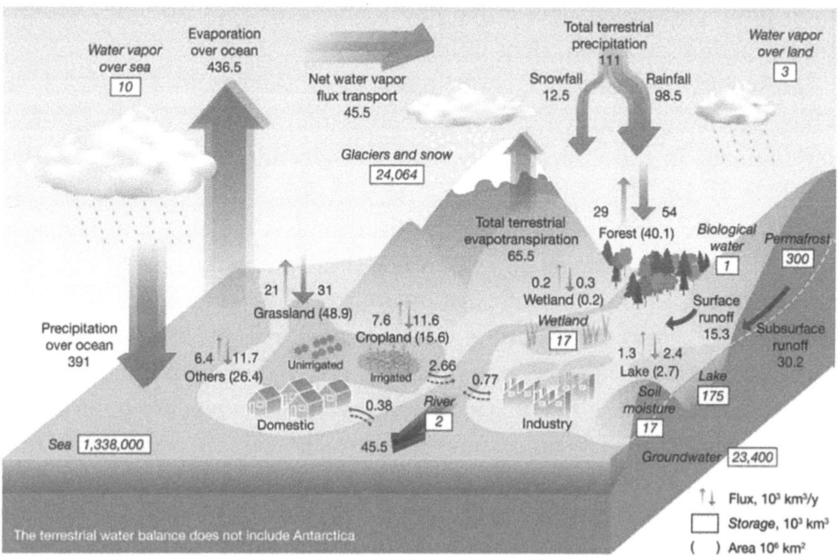

Figure 14.2 The world water availability and usage, after OkiT (2006).

to river flow in 4 min time only. Therefore, one can see that the world is not running out of water, but water is running out of rivers. It also highlighted that in future, the scientific community needs to redefine our water resources by covering the run-off to the sea.

The first author who attended the UN-Water workshop in Italy on 18–20 September 2019 proposed that:

(1) Floodwater to the sea is one of water resources together with surface/groundwater;

(2) water price is determined by construction cost; water's value is determined by the gap between water demand and water supply; and

(3) the most valuable water appears in coastline relative to the remote mountainous areas.

It is expected that the hydrologic community supports the new scientific definition of water resources and the UN-Water would issue guide the world to develop run-off lost to the sea in practice.

14.2 From monitoring to data and from water to storage

For any coastal reservoir (CR) development, without any doubt, the data collection is the first challenge. The data required are similar to that required for dam's design, including hydrological information or extreme flood flow, climate change, urban development or land use, run-off estimation, sediment and nutrient influx etc.

However, the difference is obvious as demonstrated in Table 14.1, in which the dam construction is identified as the mode of water resources development in the 20th century, and CRs are the new mode of water resources in the 21st century.

To obtain sufficient good-quality data for CR's design and management, the data from satellites, drones and probes/sensors for onsite measurement are the key information sources. The parameters should at least include clouds, temperature, winds, rainfall, run-off, land use and the parameters of water quality, e.g. nutrients, salt, turbidity and chemical/oil spill from factories/ships. The parameters should also cover the information of wave heights/tidal ranges/currents, beach stability, sea level rise, hurricanes/typhoons/tsunami, etc.

The CR designers must also consider the measured parameters listed earlier using all possible means; these data generally are limited within 30-year period. The historical data must be included from the onsite investigation also. Furthermore, the future land use and hydraulic projects must be also included; among them, the impacts of urban development and agricultural fertilizer/pesticide usage cannot be underestimated for peak flow and water quality. Simply speaking, the run-off data and water quality data in the past, current and future are equally important. Therefore, the real-time measurement will be a key component for future CR development.

The rainfall-run-off/water quality data form the basis for CR's design. To achieve better outcomes, it is expected that an integral model using supercomputing could be developed in future where different scenarios of

Table 14.1 Different modes of water resources development in the 20th/21st century.

	Dam's design	Coastal reservoirs' design
Location	Upper region	Lowest region
Catchment	A small region in upstream	Whole catchment
Water resources development	In dry seasons	In flood seasons
Impact of urban development	Mild	Significant

climate changes/sea level rise can be included for comparison. Therefore, numerical modelling could play a key role for future CR development.

The data of water use are also essential for a CR's design. This includes the water use for agriculture, industry and coastal cities; globally they are 2660, 770 and 380 km^3/year, respectively. Once the CRs are constructed, the population in coastline would be increased, and downstream industrial/agricultural activities will be intensified also; thus, the water sources in the sea will nourish the coastlines. In the 20th century, the upstream water management has sustained the inland economic development; it is expected that downstream water management will propel the coastal economic blooms. The shift from upstream water management to downstream water management may yield similar effects such as the shift from supply management to demand management.

It is important to note that water drives everything; thus, the water demand should not be underestimated. The future water demand is an important parameter to determine the CR size. Normally a CR should be large enough to supply sufficient water, but an oversized CR may cause water quality—related problems such as algal blooms. Hence a CR size should match its water demand, not too big, not too small. City planners and water resources engineers should contribute to this research direction.

The CR shape is also a scientific challenge. Together with its inlets/outlets and geomorphic conditions, a CR should have minimum detention time or water age. A stagnant water body always has poorer water quality; thus, how to keep water moving in a CR becomes a big challenge. A shallow water body should be excluded as the evaporation loss may be too high in extreme dry years. This becomes hydraulic engineers' research direction.

Different from dams, a CR always has problem of seawater intrusion. Its initial seawater should be replaced by freshwater as soon as possible. After construction, a CR also faces the seawater pollution caused by overtopping seawater, seepage from the bottom and accumulated salt from its river flow. High attention must be paid to the parameter of salinity in a CR design. The specialists in groundwater should make this very clear before design and construction.

The worst enemy for a CR is its water quality. A CR can collect every drop of water from its catchment; it also has potential to catch all pollutants generated from the catchment. It is compulsory to distinguish high-quality water from a real-time monitoring system, and only the high-quality water is allowed to enter the CR for storage. Even so, wetland technology should be considered for further purification, especially for water supply CRs. In some cities with very severe water stress problem, one should consider every drop as its water resources even polluted; in such case, membrane technology fed by CR water may be applied, like Singapore's Marina Barrage. This could be a research direction for water quality engineers.

For hydrodynamics, forces on the barriers induced by waves are similar to that on breakwaters and ports. It is also a challenge for coastal engineers.

The most important side for CR development is CR's environmental impacts. Only those with minimum environmental impacts among all water solution can be accepted by the IACRR and be recommended by the IACRR to its local government. It is expected that environmental impact assessment specialists can make new guidance for CRs as it is different from dams, which has some guidance.

14.3 Water–food–energy nexus—third-generation coastal reservoir?

Water generates life including foods that people depend on. In somehow, people are a water hunter and water guides or dominates people's activities. This is why old civilizations origin from rivers such as Nile river, Tigris and Euphrates, Indus River and Yellow River, the cradle of Egyptian, Mesopotamian, Indian and Chinese civilizations, respectively. Before Industrial Revolution in 1760, people are almost uniformly distributed on the arable land, especially on river valleys that supply the needed food. The Industrial Revolution cuts off the umbilical cord between people and this inhabited land, people started to migrate to cities in large scale or urbanization has emerged. The invention of concrete and especially the invention of large dams accelerate the modern urbanization, and megacities start to appear; the acceleration becomes more apparent after the World War II. A close look at a world cities' map reveals that coastal cities' growth rate is much higher than that of inland cities. 50% of the world's population live within 200 km of the coast in 2001, and three-fourths of mega cities are situated by sea. It is predictable that in future, more coastal cities will appear, and coastal economic corridor will be formed. The drinking water source of coastal cities will come from CRs. In the future, the coastal population could be enhanced to 60%–80%, and the sufficient water supply from CRs can help to achieve SDG6; as the vast inland area has only 30%–40% of total population, the water demand can be met easily relative to the coastal communities.

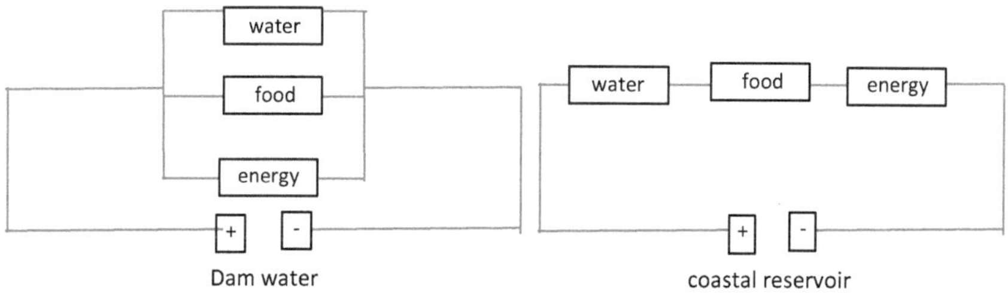

Figure 14.3 Upstream dam water management and downstream CR water management. *CR*, coastal reservoir.

Currently, the water–energy–food nexus can be simplified as three electric resistances in parallel as shown in Fig. 14.3, in which the 'water' in the blocks refers to water supply.

The water–food–energy nexus means that if more water is used for cities, then water for energy and agriculture is reduced when three of them are in parallel. But this dilemma will be changed if CRs are developed that three of them are in series, because the water for hydropower is not wasted as the CRs can still collect the lost water in the sea. Similarly, the water for irrigation will become groundwater and base flow of rivers, which can be also developed by CRs. In other words, three of 'electrical resistances' are in series, and 'the current' will not be reduced and can be used by the next one. In other words, all rivers are an open system, but CRs can make them to be a "closed loop" to recycle the water frrom sea to land.

The CRs can provide an ideal working environment for solar panels, which can generate solar energy and tidal energy if special design is applied; the solar panels can also reduce evaporation loss on the other side. CRs can be also used for energy storage by changing the electric energy into potential energy if the demand for electric energy is less, and turbines can generate electric energy when the power demand is higher than the supply. It is possible that the third-generation CR may provide a solution for the nexus of water–energy–food.

14.4 Integral management of water resources, human resources and land resources along coastline

It is predictable that the construction of CRs will attract people to live and work surrounding water sources, or CRs will nourish coastal communities. The future coastal infrastructures may be affected by CRs such as ports, airports, cities, industrial parks, highways and so on. A CR may be also used for transport highway to cross a river.

In the world, almost all existing CRs are designed for a single purpose like mitigation of seawater inundation by tsunamis or hurricanes (New Orleans, United States), water supply (Shanghai, China), coastal city's land subsidence (Jakarta, Indonesia), prevention of seawater intrusion (Alexandrina Lake, Australia) and urban regeneration (Cardiff Bay, United Kingdom). In future, a multipurpose CR is needed, and one of the example is shown in Fig. 14.4, which includes high-quality water storage for drinking purpose; one of its water bodies can be used for poor-quality water storage. If the quality of incoming water is very good, the water can be stored in one of the reservoirs via its gate; otherwise, the gate is closed. If the river water is heavily polluted, we can open another gate, so the river flow is diverted into the another reservoir for treatment till the water meets government's standard before discharging into the sea. The river flow between these two extreme qualities will be discharged into the sea via the tidal gate as normal.

The tidal gate will be closed to mitigate the seawater inundation during extreme wave heights such as hurricanes or tsunami. These events are not frequent, so the tidal gates may be closed one to two times every year or even less. Therefore, the seawater inundation caused by hurricanes/tsunamis and land subsidence can be prevented or mitigated. The tidal gate can be also used for connecting both banks of the river like a bridge.

Before the arrival of peak flood waves from upstream, water in the CR can be dumped into the sea during low tides. The empty reservoir can be functioned as a giant flood detention pond; thus the city inundation by excessive rainwater can be also mitigated.

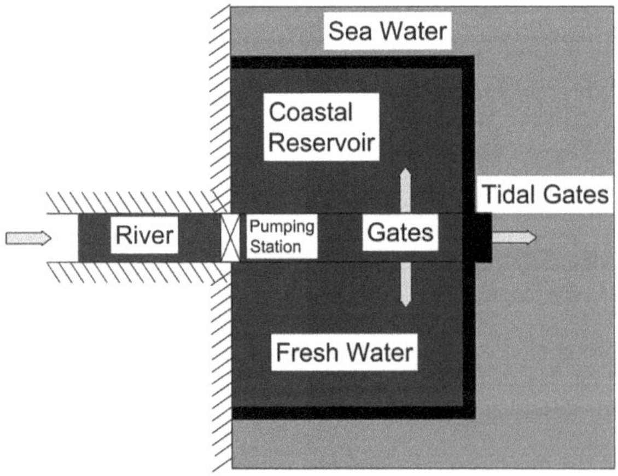

Figure 14.4 Schematic of a coastal reservoir that enables storage of freshwater during river floods.

Unlike inland dams, a CR is very flexible for its location; the suitable sites include an estuary, gulf, bay or sea, dependent on the specified problems. It has no problems such as land acquisition, land submergence and forest submergence.

These reservoirs at different locations can be interlinked through underwater subsurface pipes for transferring water from one reservoir to another in shallow seawater (herein, called as Sarovar Mala in Fig. 14.4; Sarovar means freshwater pond in Sanskrit and mala means necklace around the land) (Yang 2004, 2005; Sitharam, 2016, 2017). This is an improved 'Rivers Inter-link plan'.

14.5 Conclusions

This chapter outlines preliminarily the possible research directions for future CR development. It is certain that the research directions are different from each other in different countries/regions/river basins; it is very hard to have universal directions for every CR. The research directions suggested in the chapter include data measurement, data technology, modelling and simulation for water availability and water demand, water quality, water security and third-generation CR. The impacts of CR technology for coastal economic should not be underestimated.

References

OkiT, 2006. The University Global Hydrological Cycles and World Water Resources. Science 313, 1068. https://doi.org/10.1126/science.1128845.

Sitharam, T.G., 2016. Smart water network to store river floodwater in sea based reservoirs and a method of interlinkingthese reservoirs (Sarovar Mala). India patent no. TEMP/E-1/44294/2016-CHE.

Sitharam, T.G., 2017. Efficacy of coastal reservoirs to address India's water shortage by impounding excess river flood waters near the coast. Journal of Sustainable Urbanization. Planning and Progress 2 (2), 50–55.

Yang, S.-Q., 2004. Global and China's Water Crisis and Its Solutions in the 21st Century. Tianjin University Press, Tianjin. ISBN 7-5618-2069-0 (in Chinese).

Yang, S.-Q., 2005. "Coastal Reservoirs". Patent No. 200504653-7. (in Singapore).

Index

Note: 'Page numbers followed by "t" indicate tables, "f" indicate figures and "b" indicate boxes'.

T

Y

Z

Kolibri 360 GmbH
Ferdinand-Jühlke-Straße 7
99095 Erfurt, Deutschland
produktsicherheit@kolibri360.de